14

LITHIUM AND ANIMAL BEHAVIOR

LITHIUM AND ANIMAL BEHAVIOR
Volume II

Donald F. Smith, Ph.D.

Psychiatric Hospital
Risskov, Denmark

Lithium Research Review Series

Series Editor: **Dr. David F. Horrobin**

HUMAN SCIENCES PRESS, INC.
72 FIFTH AVENUE,
NEW YORK, N.Y. 10011

Printed in the United States of America
23456789 987654321

Library of Congress Cataloging in Publication Data   ·

Smith, Donald F., 1945–
  Lithium and animal behavior.

  (Lithium research review series)
  Bibliography: v. 2, p. 105
  Includes indexes.
  1. Lithium—Physiological effect. 2. Animals,
Habits and behavior of. I. Title. II. Series. [DNLM:
1. Ethology—Drug effects—Periodicals. 2. Lithium—
Periodicals. WI LI1821N 03]
QP535.L5S64   615'.7882   LC 81-13321

ISBN 0–89885–075–4 (v. 2)   AACR2

CONTENTS

FOREWORD

My aim in this review is to give a concise, comprehensive account of the effects of lithium on behavior of animals. The term "lithium" is used throughout this review to mean lithium ion or lithium salt. The findings mentioned are selected from Schou's bibliographies (1978; 1979a; 1980) and from information stored in the Lithium Information Center (Greist, Jefferson, Combs, Schou and Thomas, 1977). Most of the studies mentioned either appeared in journals published after completion of the first volume of this series (Smith, 1977) or failed to be considered in that volume. Authors of reports on lithium and animal behavior are requested to draw my attention to their future publications so that they can be considered for subsequent reviews in this series.

Donald F. Smith, Ph.D.
Psychopharmacology Research Unit
Psychiatric Hospital
Risskov, Denmark

Chapter 1

INTRODUCTION

Current interest in effects of lithium on behavior of animals stems mainly from two sources.
One source is the use of lithium to treat mental
disorders. The other source is the use of lithium to study laws of learning.

Topics such as alcoholism, aggression, mania and
depression are considered in animal studies related to uses of lithium in human mental disorders. The studies on alcoholism usually involve
changes in the intake of an ethanol solution
offered to animals treated with lithium. The
studies on aggression concern primarily reduction of fighting in animals given lithium. The
studies on mania and depression typically use
animal models to explore possible causes of
lithium's antimanic and antidepressant effects
in humans. The animal models used most often
are normal activity rhythms and abnormal behavioral states induced by certain drugs.

Most reports using lithium to study laws of learning concerned conditioned aversions. An upsurge in reports on lithium-induced conditioned aversion occurred in recent years, so more information on this topic appears in this book than in the previous one. Two major topics considered were the prevention of animal predation and the role of stimulus variables in the formation of conditioned aversions. A central issue of these studies is whether all aspects of conditioned aversions can be accounted for fully by traditional laws of learning (Seligman, 1970; Rozin and Kalat, 1971; Logue, 1979). This literature is included because, in addition to telling about learning, it provides information about lithium's effects on behavior of animals under certain circumstances. Of course, those who lack interest in the use of lithium in animal research on conditioned aversions are welcome to refrain from reading the sections on that topic.

It is of interest to compare the design of studies carried out either to understand lithium's actions on mental disorders or to investigate laws of learning. A basic difference between these two types of studies concerns the lithium treatments. Advantage is taken of the toxic effects of lithium in most studies on laws of learning. In these studies, lithium is used as a poison, and the animals typically receive a single dose of lithium in the range known to produce prompt signs of malaise. In contrast, attempts usually are made to minimize toxic effects in animal studies presumably related to psychiatric uses of lithium. In these studies, animals typically receive short-term (2-6 days)

or long-term treatment with lithium at doses
that fail to produce obvious signs of sickness.
A clear distinction between toxic and nontoxic
lithium treatments is often difficult to make,
however, so it is likely that animals experience
adverse effects of lithium at the time of tests
even in some studies designed to examine psychiatric uses of lithium. Consequently, it is
usually worthwhile to keep toxic effects of lithium in mind, particularly when considering
behavioral actions abtained with lithium treatments known to cause conditioned aversions.

The three main variables in studies on effects
of lithium on animal behavior are the species,
behavioral test and lithium dosage used. This
book follows the same format used in my previous
book (Smith, 1977) with respect to these variables. I use a species-specific approach and
describe the effects of lithium on behavior in
each species separately. I give attention to
the method by which the hehavioral measurements
are made in the studies cited. I pay particular
attention to the details of the lithium treatment
used in the studies cited and convert all lithium dosages to mEq lithium ion per kg body weight
in order to ease comparisons between studies.
Unfortunately, some reports lacked sufficient
information for me to determine the lithium
dosage exactly. In these cases, I either wrote
to the author to obtain the information or made
certain assumptions concerning variables such as
the body weight of the animals used and the
amount of food and fluid they consumed. I re-
quest the inclusion of sufficient information in
all future reports to enable lithium dosage to
be determined readily.

Chapter 2

EFFECTS OF LITHIUM ON INVERTEBRATES

EFFECT OF LITHIUM ON RHYTHMIC MOVEMENT IN MARINE INVERTEBRATES

Hoffmann and Smith (1979) studied the effect of lithium on movement of jellyfish (*Aurelia aurita*). The jellyfish swam either singly or in pairs in sea water maintained at 17°C. They showed an average of 10-30 uniform contractions per minute under baseline conditions. Addition of lithium to the sea water at concentrations up to 7 mmol/l led to an increase in the frequency of contractions of the jellyfish. Then, the frequency of contractions tended to decrease as the concentration of lithium increased further to 30 mmol/l. The authors attributed the decline in the frequency of contractions seen at lithium concentrations above 6 mmol/l to lithium-induced depolarisation of postsynaptic membranes, but gave no explanation for the rise in frequency of contractions seen at lower concentrations. The only other published report on behavior of invertebrates exposed to lithium describes effects on the movement of marine *limax* amoeba (Pantin, 1926). Under baseline conditions, the amoeba moved at an average rate of 2.15 μ per second in sea water. They failed to move in 0.4 M $CaCl_2$, but moved in a solution of $CaCl_2$ + LiCl. Their rate of movement depended on the ratio between the amount of lithium and calcium in the solution. The maximum rate of movement achieved was 0.54 μ per second in medium with a lithium to calcium ratio of 5, which corresponds to a lithium concentration of 0.33 M and a calcium concentration of 0.07 M. The rate was less at either higher or lower lithium:calcium molecular ratios. No movement occurred in solutions containing either 0.25 M LiCl + 0.15 M $CaCl_2$ or 0.39 M LiCl + 0.01 M $CaCl_2$. Pantin considered amoeboid movement to be similar to rhythmic contractility, in that the protoplasma of a moving amoeba undergoes a rhythmic change of state from sol to gel. The maintenance of membrane permeability and ectoplasmic viscosity appeared to be necessary for amoeboid movement to occur.

The lack of information on effects of lithium on behavior of marine invertebrates deserves comment because these animals are well-suited for study. The composition of their natural habitat, sea water, can be manipulated easily and gradually by addition of lithium or by substitution of lithium for other ions. The relatively simple organization of the nervous system of some marine invertebrates and the relatively direct relation between neuronal activity and behavior in some species may be of use for studies on lithium. It is of interest, for example, to know whether actions of lithium depend on either the stage of

development of the structure of the nervous system. Comparative studies on behavioral effects
of lithium in species of marine invertebrates
may provide information on basic mechanisms of
action of lithium on behavior.

EFFECT OF LITHIUM ON CIRCADIAN RHYTHM
IN COCKROACHES

Hofmann, Günderoth-Palmovski, Wiedermann and
Engelmann (1978) examined effects of lithium on
the circadian rhythm of male cockroaches (*Leu-
cophaea maderae*). The cockroaches lived in activity wheels with continuous red light illumi-
nation. They received a solution of lithium
chloride to drink for about 1 hour daily, with
no other source of fluid for 30 days. The concentration of the lithium solution was either
10 mmol/l, 50 mmol/l or 100 mmol/l. The length
of the circadian rhythm increased in most of the
cockroaches soon after lithium treatment began.
The increase was most marked in cockroaches
given the most concentrated lithium solution.
A few cockroaches showed a delayed response to
lithium, in that their circadian period increased about 3 weeks after the start of treatment.
More than half of the cockroaches died during
lithium treatment. No explanation is given to
account for the period lengthening effects of
lithium on the circadian rhythm of the cockroaches.

Chapter 3

EFFECTS OF LITHIUM ON FISH

EFFECT OF LITHIUM ON LEARNED FOOD AVERSION IN FISH

Mackay (1974) used lithium to study learning in Atlantic cod (*Gadus morhua* L.). The fish lived in large tanks of sea water and learned to eat pieces of liver, squid or weiner held under the surface of the water. The cod received an ip injection of lithium at a dose of either 11.9 or 15.1 mEq/kg soon after eating a food on one day. The high lithium dose caused some fish to regurgitate. Fish given the high lithium dose showed a general reluctance to eat food offered 2 days later. The novelty of the food influenced the behavior of the fish. Their reluctance to eat was greater and more prolonged for the same novel food eaten just before the lithium injection than for other foods. The findings added Atlantic cod to the list of animals that learn food aversions mediated by lithium toxicity (Gustavson, 1977). The occurrence of poison-induced learning in animals as low on the evolutionary scale as fish suggests that such

15

learning may be a basic associative process. Studies on lithium-induced food aversions in even lower animals, such as invertebrates, may provide further information on this issue.

EFFECT OF LITHIUM ON SHOALING BEHAVIOR IN FISH

Johnson (1978; 1979a) studied effects of lithium on shoaling behavior of goldfish (*Carassius auratus*). The goldfish swam for 6 hours in a solution of lithium chloride at concentrations of 10-30 mmol/l prior to the tests. They swam in a glass tank containing dechlorinated tap water during tests, either singly, in pairs, or in groups of three. Each test took 4-5 minutes. The location of the fish with respect to each other or the distance between fish served to measure shoaling. A higher score for shoaling was obtained by fish close together than by fish far apart. Lithium reduced shoaling in pairs of fish, provided that both swam in lithium before the test. The effects of lithium on shoaling in groups of three fish depended on the number of fish exposed to lithium before tests; less shoaling occurred in groups with two lithium-pretreated fish than in groups with one lithium-pretreated fish. Johnson offered three explanations to account for the effects of lithium on shoaling of goldfish. One was that the lithium treatment disorientated the fish so that they swam erratically. Another was that lithium expanded the personal space of the goldfish so that they maintained effective social contact even though they were far apart. The third was that lithium reduced the significance that the fish attached to stimuli emitted by other fish.

EFFECT OF LITHIUM IN FISH SWIMMING
IN A MAZE

Johnson (1979b; 1979c) examined effects of lithium on the behavior of goldfish in a Y-shaped
maze. The color of the walls of each arm of the
maze could be changed from white to black or an
intermediate shade of grey. The fish swam for
4-7 days before tests in a solution of lithium
at a concentration of 10-15 mmol/l. At the
start of tests, all three arms of the maze were
the same color, either white or black. Next,
one arm of the maze was made "novel" by chang-
ing its color. The fish normally swam into the
novel colored arm when the color of one arm was
changed from white to black or from black to
white. Lithium failed to affect the tendency
of fish to swim into the novel black or white
arm. However, lithium reduced their tendency
to swim into novel grey arms. Evidently, the
effects of lithium on the behavior of the goldfish depended on sensory stimulation. Johnson
explained the findings by suggesting that lithium affects stimulus generalization and the processing of stimuli around sensory treshold levels.
Johnson (1980) Studied also effects of lithium on
swimming of goldfish in a maze with several choice
points. The fish swam in water containing lithium (10 mmol/l) for 2-9 days before tests. The
lithium treatment influenced the tendency of
goldfish to turn at choice points. Evidently,
lithium caused the fish to swim erratically.

Chapter 4

EFFECTS OF LITHIUM ON BIRDS

CONDITIONED AVERSION INDUCED BY LITHIUM IN BIRDS

Interest in effects of lithium in birds stems mainly from research on learning processes. The main issue concerns the use birds make of visual and gustatory cues in feeding (Garcia and Hankins, 1977). Brett, Hankins and Garcia (1976) used lithium to condition aversions in hungry captive hawks (*Buteo jamaicensis* and *Buteo lagopus*). In particular, they examined the ease with which hawks acquire conditioned aversions based on visual and taste cues. The visual cue was the natural coat-color of the mice (white or black) and the taste cue was quinine. They measured conditioned aversions in terms of the time taken by hawks to attack and to consume mice. Before lithium treatment, the hawks quickly attacked and consumed live or dead mice thrust into their cage. The hawks received lithium (5.1 - 7.6 mEq/kg) after they attacked and ate colored

and/or flavored mice. The hawks rapidly acquired a conditioned aversion to attacking black-
quinine mice. The hawks required several lithium treatments to form a conditioned aversion to
attacking black, normally-tasting mice. A conditioned aversion to attacking mice failed to
occur on the basis of taste cues in the absence
of distinct visual cues, although hawks did even-
tually acquire a conditioned aversion to eating
quinine-tasting mice. The findings show the
consummatory behavior of hawks to be governed
mainly by taste cues, while visual cues can
influence attack behavior (Garcia, Rusiniak and
Brett, 1977).

Gustavson (1977) reported lithium-induced conditioned aversions in two magpies (*Pica pica*).
The birds received lithium in pieces of rabbit
meat. Consumption of the lithium-treated meat
caused the birds to vomit. Subsequently, they
refrained from eating rabbit meat. The behavior
of the magpies suggested that they first acquired
an aversion to the flavor of meat and that the
flavor aversion mediated an aversion based on
visual cues.

Gaston (1977) used lithium to condition aversions
in young domestic Leghorn cockerels. The chicks
received lithium (1-2 mEq/kg) after drinking a
solution flavored with sucrose and/or colored
with green vegetable dye. Chickens given lithium after drinking sweet-green solution showed
an aversion to the solution subsequently. Gaston
obtained inconclusive findings in chicks given
lithium after drinking either a sucrose flavored
solution or the green colored solution. The
chicks failed to acquire reliable conditioned

aversions based on taste cues or color cues alone.
Nevertheless, the results show chickens to be
capable of 1-trail lithium-induced conditioned
aversion learning.   Evidently, specialized learning ability is present in the relatively simple
nervous system of chickens shortly after birth.

Martin, Bellingham and Storlien (1977) also used
yound chickens to study conditioned aversions
induced by lithium.   Their chickens received lithium (6-8 mEq/kg) soon after eating a novel
food.   They found the lithium treatment to produce conditioned food aversions based on either
color or texture.   The amount of previous experience the chicks had with the novel stimuli influenced the strength of the lithium-induced conditioned aversions.   Martin et al. interpreted
their findings in terms of the predisposition of
animals to associated eating-related cues with
sickness.

EFFECT OF LITHIUM ON WATER INTAKE
OF BIRDS

Westbrook, Hardy and Faulks (1979) observed increased water intake in pigeons given lithium
(0.85 - 3 mEq/kg).   The pigeons were first deprived of water, then given water to drink, next
given lithium by injection and finally given
water to drink again.   The pigeons began to drink
enhanced amounts of water within 1-3 hours after
administration of lithium and continued to show
increased water consumption for 2-3 days.   The
effect of lithium on water intake was dose-dependent.   Although the exact mechanisms of lithium's
action were unknown, Westbrook et al. postulated
lithium to exert its effects centrally.   They
viewed the enhanced water intake in pigeons given

lithium as part of a general response to poisoning.

BEHAVIOR OF LITHIUM-INTOXICATED BIRDS

Vomiting and head-flipping occurred in hawks soon after an injection of lithium (5.1 -7.6 mEq /kg) (Brett, Hankins and Garcia, 1976). Lethargy and diarrhea appeared in chicks given lithium (1-2 mEq/kg) by injection (Gaston, 1977). Pigeons showed diarrhea, shivering and vomiting after an injection of lithium (0.85 - 3 mEq/kg) (Westbrook, Hardy, and Faulks, 1979). Downie, Wasnidge, Floto and Robinson (1977) found female Japanese quail given lithium (7.5 mEq/kg) to lay eggs with thin shells or no shells at all. Lithium (7.5 - 15 mEq/kg) killed some male Japanese quails within 24 hours. Kidney failure appeared to contribute to the toxic actions of lithium in quail.

Chapter 5

EFFECTS OF LITHIUM ON MICE

EFFECTS OF LITHIUM ON LOCOMOTOR ACTIVITY OF MICE

The types of locomotion examined in mice given lithium were circadian activity, spontaneous activity, exploratory behavior, motor coordination and curiosity-related movement. It is to be noted that distinctions between these types of activity are often difficult to make and are usually a matter-of-opinion of those who carry out the experiments. They serve, nevertheless, to categorize effects of lithium on locomotor activity.

Poirel, Hengartner and Briand (1979) studied effects of lithium on the circadian rhythm of locomotor activity of mice. The mice received a lithium solution to drink and consumed a daily lithium dosage of about 0.45 mEq/kg. Lithium failed to affect the circadian rhythm of the mice tested for 15 minutes in a circular open field at times during the day and night. Some other researchers measured spontaneous activity of mice in photocell cages (Bignami, Pinto-Scognamiglio and Gatti, 1974; Borison, Sabelli, Maple, Havdala and Diamond, 1978; Wielosz, 1979). They found no effect of lithium at daily doses of 1.1 - 5 mEq/kg given for 1-8 days before activity tests. Bignami et al. measured also exploratory behavior as the number of head-dips shown by mice on a holeboard. Lithium, at a dose of 2-5 mEq/kg, decreased exploratory behavior of mice on some days but not on others. Bignami et al. stated that further experiments showed the effects of lithium on exploratory behavior to vary from one experiment to another without apparent reason. They concluded that lithium probably has few or no specific effects on activity and exploration of mice, a conclusion in agreement with previous findings (see Smith, 1977).

Malick (1978) checked effects of lithium on motor coordination of mice in a rotarod test. Mice used in the study were able to stay on the rotating rod for one minute before lithium treatment. Five days of lithium injected at a dose of 2.4 - 7.1 mEq/kg failed to affect the rotarod performance of the mice.

Weischer (1979) tested mice in an octagonal enclosure with small holes in the walls. The number of times a mouse looked through a hole served to measure curiosity. Mice drank a solution of lithium for three weeks before tests. The daily lithium dosage consumed was around 12 mEq/kg. Some of the mice lived alone, in isolation, throughout the experiment. The lithium

treatment restored the curiosity of isolated
mice to normal levels. In addition, Weischer
found lithium to increase the motility (hori-
zontal movement) of isolated mice, but not to
affect their rearing (vertical movement). Weischer concluded that lithium had some activating
effects on the locomotor hehavior of mice.
Poirel, Hengartner and Briand (1979) also noted
an activating effect of lithium on a locomotor
activity of mice. They studied the circadian
rhythm of grooming behavior in mice given lithium to drink at a daily dose around 0.45 mEq/kg.
Lithium enhanced grooming during the day and
night. Evidently, lithium influences only certain types of activity of mice.

EFFECT OF LITHIUM ON BEHAVIORAL
ACTIONS OF AMPHETAMINE AND RELATED
DRUGS IN MICE

Interest in effects of lithium on behavior of
mice given amphetamine stems mainly from the
notion that actions of amphetamine may provide
an animal model for mania (see Murphy, 1977).
Several recent studies show effects of lithium
on hehaviors seen in mice given amphetamine or
one of its derivatives. As in previous reports
(see Smith, 1977), lithium antagonized behavioral effects of amphetamine in some experiments
and enhanced them in others. Borison, Sabelli,
Maple, Havdala and Diamond (1978) gave mice lithium for 8 days at a daily dose of 1.1 mEq/kg.
This treatment reduced Straub tail and vocalization induced by amphetamine in mice, but failed
to affect piloerection, stereo-typic movements
and jumping. Borison et al. related the effects
of lithium to antagonism of actions of amphet-

amine on the synthesis and disposition of phenylethylamine in brain. Berggren, Tallstedt,
Ahlenius and Engel (1978) gave groups of mice
lithium (4.05 - 8.1 mEq/kg) together with d-
amphetamine and measured their activity automa-
tically with photocells. The stimulation of
locomotor activity produced by amphetamine was
suppressed by lithium in a dose-dependent fash-
ion. Administration of l-dopa plus Ro-4-4602
antagonized the suppressant effects of lithium
on the locomotor stimulation induced by amphetamine. Berggren et al. observed no effect of
lithium on the behavioral stimulation induced in
mice by apomorphine or clonidine. They concluded that inhibitory effects of lithium on locomotor stimulation induced by amphetamine in mice
are probably mediated by actions on presynaptic
catecholaminergic neurotransmission.

Others found lithium to potentiate effects of
amphetamine. Abdallah, Riley, Boeckler and
White (1977) noted an increase in circling mo-
tion, gnawing and licking in mice given d-amphet-
amine after consuming lithium in their diet at a
daily dose of 5-10 mEq/kg for 12 days. Yamada
and Furakawa (1979) described a behavioral syndrome of tremor, rigidity, Straub tail, hind-
limb abduction and reciprocal forepaw treading
("paino-playing") in mice given p-chloroamphet-
amine together with lithium (10-25 mEq/kg. They
considered the syndrome to reflect overstimula-
tion of postsynaptic serotonergic receptors.
Wielosz (1979) studied effects of lithium on the
toxicity of dl-amphetamine in groups of mice.
The mice received lithium (1.2 - 4.8 mEq/kg) by
injection for four days. The treatment increased the lethality of amphetamine. Janowsky,

Abrams, Groom, Judd and Cloptin (1979) examined
the effect of lithium in mice given methylphenidate. The mice received lithium at a daily
dose of 2.4 mEq/kg for seven days before tests.
This lithium treatment failed to affect the
stereotypic gnawing induced by methylphenidate
although an intensification of methylphenidate-
induced gnawing occurred in mice given lithium
either acutely or in higher doses. In addition,
the lithium treatment antagonized the suppressant effects of physostigmine on methylphenidate-
induced gnawing. Janowsky et al. suggested that
lithium may potentiate effects of methylphenidate on behavior, decrease entrance of physostigmine into brain or decrease cholinergic neurotransmission.

EFFECTS OF LITHIUM ON HEAD-TWITCHES
IN MICE

Head-twitching in mice is a behavior related
primarily to activation of serotonergic neurotransmission. Yamada and Furakawa (1979) reported effects of lithium on head-twitching in
mice. Although administration of lithium alone
at a dose of 10-25 mEq/kg failed to produce head-
twitches, they occurred in mice given lithium
together with either reserpine, tetrabenazine or
syrosingopine. Lithium also enhanced head-
twitches produced by 5-methoxy-N,N-dimethyl
tryptamine. Further pharmacological studies
showed the head-twitches induced by lithium in
combination with other drugs to be due mainly
to release of serotonin and potentiation of its
postsynaptic actions (Yamada and Furakawa, 1979;
Furakawa, Yamada, Kohno and Nagasaki, 1979).
Noradrenergic and cholinergic neurotransmission

appeared to play a minor role in the effects of
lithium on head-twitches in mice.

EFFECT OF LITHIUM ON MORPHINE ANALGESIA
AND DEPENDENCE IN MICE

Bulaev and Ostrovskaya (1978) examined effects
of lithium on analgesic actions of morphine in
mice. They measured both the pain threshold and
the duration of analgesia produced by morphine.
The criterion for pain threshold was the production of prolonged vocalization in response to
electrical stimulation of the tail in mice.
Bulaev and Ostrovskaya state that lithium weak-
ened analgesic actions of morphine by reducing
both pain threshold and duration of anesthesia.
Unfortunately, the translation of their original
report lacks clear definition of the symbols
used in their figure and fails to specify their
lithium dosage. Bulaev and Ostrovskaya reported
also on effects of lithium on morphine preference in mice. They made mice dependent on
morphine solutions by increasing the concentration from 0.3 to 1 mq/kg over two weeks. Mice
then received an injection of lithium (5 mEq/kg)
followed by five days of testing with a morphine
solution and water to drink. The lithium treatment reduced the preference of mice for morphine
for 3-4 days. Their results indicate that lithium has antimorphine actions, but the mechanisms responsible for the effects of lithium are
unresolved.

EFFECT OF LITHIUM ON ETHANOL
INTAKE OF MICE

Nottage, Syme and Syme (1978) carried out three

experiments on ethanol intake of mice given lithium. They took advantage of genetic differences and selected mouse strains with either a
high preference or a low preference for ethanol.
In the first experiment, they gave mice water
and an ethanol solution to drink and injected
lithium (0.3 mEq/kg) for four days. In the second experiment, they offered mice with a high
preference for ethanol two ethanol solution
(5% and 12%) to drink and gave them daily lithium injections (3 mEq/kg) for four days. In
the third experiment, mice with a high preference for ethanol received water and a 5% solution of ethanol to drink and lithium (0.3 mEq
/kg) by injection on each of four days. Lithium failed to affect ethanol intake of mice in
all of the experiments.

EFFECT OF LITHIUM ON DRUG-INDUCED
SEIZURES AND NARCOSIS IN MICE

Most studies fail to find effects of lithium on
seizures in mice (see Smith, 1977). Nevertheless, Nagai and Kwaki (1976) noted anticonvulsant actions of lithium. They induced seizures
in mice by an injection of nicotine. Mice received a solution containing lithium to drink at
a concentration that produced lithium levels in
plasma of 0.7 mEq/1 after 8-10 weeks of treatment. The lithium treatment delayed the onset
of seizures induced by nicotine and increased
the dose of nicotine needed to induce clonic and
tonic convulsions in the mice. Nagai and Kwaki
considered calcium balance and dopamine metabolism to be involved in the occurrence of seizures.

Messiha (1976) studied effects of lithium on

narcosis induced by ethanol in mice. The duration of narcosis was the time between loss of
righting reflex and its reappearance. Mice received lithium (5 mEq/kg) by injection for up
to seven days before administration of ethanol.
Up to five days of lithium treatment failed to
affect the duration of narcosis induced by ethanol in the mice, while seven days of treatment
with lithium increased the duration of ethanol
narcosis. Messiha considered the results to
suggest enhancement of ethanol toxicity by lithium. Ho and Ho (1979) also found lithium (4-
6 mEq/kg) to potentiate the duration of narcosis
induced by ethanol in mice. In addition, they
found naloxone to antagonize effects of lithium
on ethanol narcosis. Ho and Ho suggested the
effects to be related to alterations of electro-
lyte and water balance.

EFFECT OF LITHIUM ON AGGRESSION IN MICE

Lithium influenced isolation-induced aggression,
predatory aggression and fighting induced by
amphetamine, but failed to affect either spontaneous aggression or maternal aggression in
mice.

Isolation-induced aggression refers to the enhanced fighting seen in male mice placed together after living alone for some time. Eichelman, Seagraves and Barchas (1977) were the first
to report effects of lithium on isolation-induced aggression in mice. Their measure of
aggression was the amount of time the mice spent
fighting during a 15-minute test. They found
lithium at daily doses of 4.5 - 6 mEq/kg to decrease fighting in pairs of mice placed together
after 14 days of treatment. Malick (1978) en-

hanced aggression in mice by four weeks of isolation. Mice selected for the study fought
consistently when tested in pairs. Malick measured fighting as an all-or-none response in
pairs of mice. Lithium inhibited fighting after
five days of treatment. The ED_{50} for inhibition
of fighting by lithium was 1.93 mEq/kg. Malick
suggested that effects of lithium on isolation-
induced fighting of mice may be due to actions
on monoaminergic neurotransmission. Brain and
Al-Maliki (1979) also studied effects of lithium
on isolation-induced aggression. They placed
an anosmic male mouse in the home cage of an
isolated male mouse given 10-12 mEq/kg lithium
for 4-10 days. Lithium increased the latency
to attack, decreased the number of attacks and
reduced the time spent attacking by the isolated
mouse during a 10-minute test. Brain and Al-
Maliki considered lithium to reduce the motivation for fighting in isolated mice.

Predatory aggression occurs in mice offered a
live cricket (*Acheta domesticus*) or locust
(*Locustus migratorius*). The effect of lithium
on predatory aggression of mice depended on the
time of the lithium injection as well as on the
lithium dosage. Brain and Al-Maliki (1979) injected mice with lithium at a daily dose of
about 10 mEq/kg for four days *before* tests of
locust-killing. Under these conditions, lithium failed to affect predatory aggression reliably. On the other hand, Lowe and O'Boyle
(1976) and Klunder and O'Boyle (1979) found lithium injection given *after* pre atory aggression
of mice to suppress some aspects of the behavior.
Lithium given at a dose of 0.75 mEq/kg to mice
soon after attacking and eating part of a crick-

et reduced subsequent cricket eating but failed
to affect cricket-killing reliably. Lithium
given at a dose of 3 mEq/kg to mice soon after
either attacking or attacking and eating part
of a cricket reduced subsequent attack behavior
as well as cricket-eating. The influence of the
time of the lithium injection with respect to
the behavior of the mice suggests that the punishment produced by the aversive properties of
lithium mediated the reduction in predatory
aggression.

Borison, Sabelli, Maple, Havdala and Diamond
(1978) reported a preventive effect of eight
days of lithium treatment at a daily dose of
1.1 mEq/kg on aggressive fighting in groups of
mice given *d*-amphetamine.

Weischer (1979) gave lithium to groups of mice
in their drinking water for three weeks. The
daily lithium dose consumed was about 12 mEq/
kg. Weischer noted no consistent effect of this
lithium treatment on the spontaneous aggressive
behavior of the mice.

Brain and Al-Maliki (1979) studied the maternal
aggression shown by lactating female mice toward an anosmic male mouse "intruder". The
female mice received lithium at a daily dose of
about 10 mEq/kg for 4-10 days. The lithium
treatment failed to affect the latency of attack, the amount of time spent attacking or the
number of attacks of the female mice on the intruder. Brain and Al-Maliki called attention
to the differences between the effects of lithium on several types of aggression and warned
against extrapolating too freely between dif-

ferent forms of aggression.

LITHIUM TOXICITY IN MICE

The toxicity of a single injection of lithium
in mice appears to depend on genetic factors,
route of administration, concurrent drug treatment and time of administration. Smith (1978b)
found the LD_{50} for a subcutaneous injection of
lithium to vary between 17.4 and 19.4 mEq/kg in
four strains of mice. The mice that died of
lithium intoxication usually were prostrate for
0.25 - 6 hours after treatment. Some intoxicat-
ed mice lived longer than 24 hours. They usually became emaciated and tremulous prior to
death. Ho and Ho (1978) obtained an LD_{50} of
10.3 mEq/kg in Swiss Webster mice given an ip
injection of lithium and found administration
of ethanol to the mice to lower the LD_{50} of lithium. Their observation is consistent with
the suggestion of Messiha (1976) concerning
toxic interaction of lithium and ethanol. Hawkins, Kripke and Janowsky (1978) noted circadian variation in lithium toxicity in mice.
They injected mice with a single, fixed dose of
lithium (22.4 mEq/kg) at times during the day
or night. The toxicity of lithium was greater
in mice injected during the day than during the
night. Hawkins et al. speculated that the
circadian variation in lithium toxicity may relate to circadian variations in other variables
such as activity, water and food consumption,
lithium metabolism and clearance, or plasma al-
dosterone.

The toxicity of **chronic** lithium treatment in
mice failed to depend on the anion injected to-

gether with lithium. Jernigan, Schrank and Kraus (1978) gave mice daily injections of lithium either as the chloride salt (12.9 mEq/kg) or as carbamyl phosphate salt (13.8 mEq/kg). Approximately half the mice died during 21 days of treatment, regardless of the lithium salt injected. Perez-Cruet and Dancey (1977) observed deaths in mice given lithium at a daily dose of 18 mEq/kg for four days. The mice showed neuromuscular irritability, polydipsia, ruffled hair and diarrhea prior to death. Lazarus and Muston (1978) gave mice lithium in their drinking water at concentrations up to 28 mmol/l. The amount of lithium consumed daily by the mice was probably 5-11 mEq/kg. Loss of body weight and death occurred during eight weeks of treatment in some of the mice given the solution with the highest lithium concentration.

Chapter 6

EFFECTS OF LITHIUM ON RATS

EFFECT OF LITHIUM ON LOCOMOTOR
ACTIVITY IN RATS

Previous research showed lithium to affect locomotor activity in rats tested in an open field
and in a relatively small cylindrical chanber,
but not in most photocell activity recorders or
in the Animex (see Murphy, 1977; Smith, 1977).
Most recent studies found no consistent effects
of lithium (0.6 - 3.6 mEq/kg) on locomotor activity in rats tested in automatic activity recorders (Flemenbaum, 1977a; Alexander and Alexander,
1978; Britton, Bianchine and Greenberg, 1976;
Wielosz, 1979; Rastogi and Singhal, 1977a; 1977b),
although Mukherjee, Bailey and Pradhan (1977)
found the effect of lithium on locomotor activity in rats to depend on the time of testing
with respect to the lithium injection. They
placed rats in an automatic activity recorder at
times after an injection of lithium (1-3 mEq/kg)
and observed the greatest suppression of activi-

34

ty 8 hours after injection. A correlation between the activity-suppressant effect of lithium and the concentration of lithium in the
brain suggested the reduction in locomotor activity to be due to actions of lithium in the central nervous system.

Lieberman, Alexander and Stokes (1979) compared
the effects of lithium isotopes on the locomotor activity of rats. They measured activity as
the number of times rats moved from one plate to
another in a small chamber. The rats received
either lithium-6 or lithium-7 each day for 5
days. Locomotor activity decreased more rapidly in rats given lithium-6 than in those given
lithium-7.

Harrison-Read (1978) observed the behavior of
rats placed in a Y-maze after short-term lithium
treatment (2 mEq/kg). The treatment reduced the
locomotor activity of the rats and increased the
amount of time they spent investigating a novel
object. Arnsten and Segal (1979) obtained similar findings in rats given lithium (2 mEq/kg)
and tested in a multicompartment chamber with
novel spherical stimuli. Harrison-Read (1978)
studied also the behavior of lithium-treated
rats placed on a board with holes in it and a
flashing light below. Short-term lithium treatment decreased the locomotor activity of the
rats in this test too. Administration of L-
tryptophan antagonized the effects of lithium on
locomotor activity in rats, which suggests a
role of central serotonergic mechanisms in effects of lithium on locomotor activity in rats.
Harrison-Read considered short-term treatment
with lithium to increase the sensitivity of rats

to relatively subtle novel stimuli. This notion contrasts with that of Johnson (1979d) on reduction of stimulus significance in rats given lithium.

EFFECTS OF LITHIUM ON DRUG-INDUCED HYPERACTIVITY IN RATS

Several studies examined effects of lithium on drug-induced states of hyperactivity presumed to be related to monoaminergic neurotransmission. Harrison-Read (1979) used drugs which act primarily on serotonergic mechanisms to study actions of lithium. The drugs caused behaviors such as postural tremor, forepaw treading, Straub tail, head-weaving and hypersalivation in rats. Short-term treatment with lithium (2mEq/kg) potentiated these behaviors in rats given fenfluramine. Long-term treatment with lithium potentiated behaviors induced in rats by fenfluramine, 5-hydroxytryptophan and 5-methoxy-N,N-dimethyltryptamine. Harrison-Read considered the findings to show lithium to impair intraneuronal storage of serotonin and to induce a state of supersensitivity of indoleamine receptors after long-term treatment. Flemenbaum (1977a; 1977b) studied effects of lithium on behaviors mediated mainly by catecholaminergic mechanisms. He determined the effect of lithium (2.5 - 3 mEq/kg) on increased running-wheel activity induced by several drugs in rats. Lithium counteracted running-wheel activity produced by *l*-amphetamine, methamphetamine and apomorphine, but failed to suppress reliable running-wheel activity caused by *d*-amphetamine or cocaine. Flemenbaum explained the findings in terms of presynaptic and postsynaptic actions of

lithium on catecholaminergic neurotransmission.
Rastogi and Singhal (1977a; 1977b) induced hyperactivity in neonatal rats by daily injections
of thyroid hormone. Administration of lithium
(1.62 mEq/kg) during the last days of hormone
treatment counteracted the development of hyperactivity in the rats. Biochemical studies showed
lithium to influence the metabolism of monoamines
in hormone-treated rats. Rastogi and Singhal
considered alterations in intracellular sodium
and potassium to be involved in lithium's effects
on neurotransmission and hyperactivity. Maté,
Ribas, Acobettro and Santos Ruiz (1979) recon-
firmed the occurrence of enhanced locomotor activity in rats given lithium (6 mEq/kg) together
with pargyline, an inhibitor of monoamine oxidase
(see Smith, 1977). Their rats showed anomalies
of locomotion such as muscular incoordination,
lack of smoothness of movement and turning about
the hindlegs. Later on, the rats lost their balance, showed trembling of the hindlegs and died.
Maté et al. considered the behavior produced in
rats given lithium plus pargyline to represent
a toxic syndrome rather than an authentic syndrome of hyperactivity.

EFFECT OF LITHIUM ON DRUG-INDUCED STEREOTYPIES IN RATS

Lithium influenced several types of stereotypic
behavior induced by drugs in rats. Janowsky,
Abrams, Groom, Judd and Cloptin (1979) induced
a state of gnawing, sniffing and repetitive movement in rats by administration of methylphenidate. They found lithium (2.4 mEq/kg) alone to
be without effect on the stereotypic behavior,
but to antagonize effects of physostigmine on

the behavior. Janowsky et al. Offered a tentative explanation for the effects of lithium based on actions on the balance between cholinergic and adrenergic neurotransmission. Smith (1978c) produced a state of head-nodding, head-swinging and sniffing with piloerection and salivation in rats given B-phenylethylamine. Lithium (1.5 - 2.3 mEq/kg) suppressed the stereotypic behaviors induced by B-phenylethylamine in rats. Administration of inhibitors of monoamine oxidase antagonized the suppressant effects of lithium on the stereotypy. Smith offered two explanations to account for the findings; one based on effects of lithium on monoamine oxidase and another based on actions of lithium on dopaminergic neurotransmission.

Several studies investigated effects of lithium on stereotypic behaviors induced by apomorphine in rats (Pert, Rosenblatt, Sivit, Pert and Bunney, 1978; Allikmets, Stanley and Gershon, 1979; Friedman, Dallop and Levine, 1979). Lithium treatment suppressed stereotypic behavior induced by apomorphine in rats given daily injections of lithium (2 mEq/kg) but not in those given lithium in their food (1-4 mEq/kg). Pretreatment of rats with haloperidol or reserpine enhanced stereotypies produced by apomorphine. Lithium prevented the enhancement in rats pretreated with haloperidol and potentiated stereotypies further in rats pretreated with reserpine. The findings suggest lithium to counteract the development of dopamine postsynaptic receptor supersensitivity induced by haloperidol and to enhance disruptive actions of reserpine on presynaptic dopaminergic neurotransmission.

EFFECT OF LITHIUM ON HEAD-TWITCHES
IN RATS

Lithium influenced head-twitches in rats. Harrison-Read (1978) found long-term lithium treatment at a daily dose of 2 mEq/kg to potentiate head-twitches induced by 5-methoxy-N,N-dimethyltryptamine or by carbidopa + 5-hydroxytryptophan. The findings pointed to an increase in the sensitivity of serotonin receptors as a mechanism of action of lithium. Later, Wielosz reported head-twitches in rats given lithium (3. 6 - 6.75 mEq/kg) (Wielosz, 1979; Wielosz and Kleinrok, 1979). Pretreatment of rats with reserpine, serotonin blocking agents or lithium antagonized lithium-induced head-twitches. On the other hand, lithium pretreatment potentiated head-twitches induced by 5-methoxytryptamine. Wielosz considered the findings to show lithium to have a transient indirect agonistic action on postsynaptic serotonin receptors in addition to enhancement of intraneuronal storage of serotonin and enhancement of serotonin postsynaptic receptor sensitivity. Thus, effects on serotonin neurotransmission appear to play a role in some actions of lithium on behavior of rats.

EFFECT OF LITHIUM ON DRUG-INDUCED
CATALEPSY IN RATS

Dessaigne, Scotto and Giugues (1978) examined effects of lithium on catalepsy induced by reserpine or neuroleptic drugs in rats. They measured catalepsy as the tendency of rats to remain in abnormal postures imposed by the experimenters. A single dose of lithium (11.78

mEq/kg) potentiated catalepsy induced by reserpine and all of the neuroleptics tested. Repeated daily doses of lithium (5 mEq/kg) brief-
ly potentiated catalepsy induced by most of the
neuroleptics, but not by reserpine. Dessaigne
et al. assumed catalepsy to be due to lack of
dopamine in the striatum. They explained lithium's actions on catalepsy in terms of depletion
of dopamine in the striatum and reduction of
postsynaptic actions of dopamine.

EFFECTS OF LITHIUM ON CONVULSIONS
INDUCED IN RATS

The influence of lithium on convulsions in rats
is variable. In most studies, lithium enhanced
seizures induced in rats. Toxic doses of lithium (12-18 mEq/kg) induced clonic seizures
followed by death in rats (Etevenon, Fraisse,
Guillon, Breteau and Boissier, 1971). Ho and
Tsai (1976) found no effect of lithium at a
daily dose of 2.4 mEq/kg on convulsions seen
after withdrawal of ethanol from physically
dependent rats, but a larger dose of lithium
appeared to potentiate the convulsions. Long-
term intake of low amounts of lithium (0.16 -
0.28 mEq/kg) enhanced the susceptibility of rats
to seizures induced by electroshock (Branchey,
Cavazos and Cooper, 1977). Direct administration of lithium (7.1 micromoles) in the brain
of rats potentiated seizures produced by ouabain,
perhaps by enhancement of calcium uptake into
cells (Inoue, Tsukada and Barbeau, 1977). Kadzielawa (1979) studied interactions of lithium
(3.6 mEq/kg) and anticonvulsant drugs on seizures induced by electroshock in rats. The
lithium treatment alone failed to influence the

induction of seizures by electroshock, but it
reduced the anticonvulsant properties of pheny-
toin, acetazolamide and methazolamide. The
antagonistic action of lithium against anticonvulsant drugs occurred also during long-term
treatment. Kadzielawa accounted for the reduction in potency of anticonsulsants in lithium-
treated rats in terms of actions of catechola-
mine neurotransmission. Banister, Bhakthan and
Singh (1976) examined effects of lithium (9.4
mEq/kg) on seizures induced in rats exposed to
oxygen at high pressure. They found lithium
to antagonize convulsions measured soon after
injection but to enhance convulsions later on.
The time course of lithium's effects made peripheral actions most likely to account for the
anticonvulsant effect.

EFFECTS OF LITHIUM ON SLEEP IN RATS

Electrophysiological studies showed dose-dependent effects of lithium on phases of sleep in
rats (Danguir, Nicolaidis and Perino-Martel,
1976; Etevenson, Braisse, Guillon, Breteau and
Boissier, 1971; Venkatakrishna-Bhatt and Bures,
1978a; 1978b). A dose of lithium below 1.5
mEq/kg had few consistent effects on sleep aside
from causing an increase in paradoxical sleep.
A higher dose of lithium (3 - 4.8 mEq/kg) caused
prompt reductions in paradoxical sleep during
the time in which the rats showed signs of distress. Later on, an excess of slow wave sleep
and paradoxical sleep occurred. An even higher
dose of lithium (> 5 mEq/kg) usually caused
abnormal electroencephalographic patterns with
rapid frequencies and high amplitudes, accompa-
nied by a reduction in total sleep time and in

paradoxical sleep. The effects of high doses on sleep probably reflect lithium toxicity.

EFFECT OF LITHIUM ON ABLATION-INDUCED HYPEREXCITABILITY IN RATS

Mukherjee and Pradhan (1976a) induced hyperexcitability in rats by ablation of the septum. Rats with septal lesions show an exaggerated startle reaction, hyperirritability and aggressiveness. Administration of lithium (1-3 mEq/kg) to rats with septal lesions reduced their hyperexcitability. Mukherjee and Pradhan suggested lithium's actions in septal rats to be mediated by effects on emotionality and motor activity.

EFFECT OF LITHIUM ON AGGRESSION IN RATS

A distinction is to be made between studies on aggression of rats pretreated with lithium and studies in which lithium is used primarily to punish aggressive behavior. Pretreatment of rats with lithium reduced irritable aggression produced by footshock as well as aggression induced by certain drugs, but failed to influence muricidal behavior. On the other hand, administration of lithium after muricide subsequently reduced the occurrence of some aspects of this type of predatory aggression.

Mukherjee and Pradhan (1976b) induced fighting in pairs of rats by giving them footshocks. The aggressive behavior stabilized after several days of testing before administration of lithium. Lithium, at a dose of 1-3 mEq/kg 4-48 hours before test, reduced the frequency of attacking and fighting during test sessions. The reduc-

tion of footshock-induced aggression of rats
given lithium agrees with previous reports (see
Sheard, 1977; Smith, 1977). Mukherjee and
Pradhan found lithium pretreatment also to reduce the enhancement of footshock-induced aggression seen in pairs of rats given amphetamine and
to inhibit the reduction of irritable behavior
seen in pairs of rats given scopolamine. They
considered the possibility that the effect of
lithium on irritable aggression may be due to
modification of both adrenergic and cholinergic
neurotransmission. Marini, Sheard and Kosten
(1979) carried out studies on irritable aggression in rats in order to explore the notion of
serotonergic involvement in lithium's mode of
action. Their rats received long-term treatment with lithium (3-4 mEq/kg). Lithium reduced
fighting elicited by footshock in pairs of rats.
What is more, it antagonized both the increase
in fighting produced by lysergic acid diethyl-
amide and the decrease in fighting caused by
chlorimipramine. Marini et al. mentioned
stabilization of postsynaptic response in serotonergic pathways as a possible explanation for
their findings and speculated also on an involvement of adrenergic systems in lithium's mode
of action.

Mukherjee and Pradhan (1976b) examined also
effects of lithium pretreatment on muricidal
behavior of rats. An injection of lithium (1-3
mEq/kg) given 0.5-4 hours before tests failed to
reduce mouse killing by the rats, in accordance
with previous reports (see Smith, 1977). They
postulated that lithium fails to influence very
strong instinctual and highly motivated responses
such as muricide. Allikmets, Stanley and Gershon

(1979) studied effects of lithium on aggression induced by apomorphine in pairs of rats. Apomorphine-induced aggression consisted of upright attack posturing, wrestling and vocalization. Rats received lithium at a daily dose of 2 mEq/kg for two weeks. Tests of aggression took place 5 days after discontinuation of lithium treatment. The lithium treatment failed to affect apomorphine-induced aggression, but it antagonized the enhancement of apomorphine-induced aggression otherwise seen in rats pretreated with haloperidol. Allikmets et al. related the antagonistic action of lithium to dopaminergic supersensitivity. They suggested that lithium blocked the development of supersensitivity induced by haloperidol in dopaminergic pathways.

Previous studies showed punishment of muricide with lithium to suppress some aspects of this predotary aggression (see Smith, 1977). Recent studies used lithium to examine the relation between the appetitive phase (attacking and killing) and the consummatory phase (eating) of muricide. Berg and Baenninger (1974) injected muricidal rats with lithium (3 mEq/kg). They permitted the rats to kill a mouse or to kill a mouse and eat some of it before the lithium injection. The lithium treatment suppressed mouse eating but failed to suppress mouse killing. In fact, the time required for the rats to attack and kill mice decreased on days after the lithium treatment. Rusiniak, Gustavson, Hankins, and Garcia (1976) carried out a series of experiments to examine further the relation between the appetitive and consummatory phase of predatory aggression in rats. In particular, they examined the role of flavor stimuli in predatory aggres-

sion. The rats used were known to be mouse kill-
ers. They received a dose of lithium (3.4 -4.3
mEq/kg) that produced lethargy in about 10 minutes.
In one experiment, rats received a dead mouse to
eat followed by an injection of lithium on several
days. This treatment produced an aversion to eat-
ting dead mice. The rats killed live mice readily, but did not eat them. Next, the rats received
a series of lithium injections soon after killing live mice. This treatment succeeded in sup-
pressing mouse killing. In the second experiment,
muricidal rats received mice soaked in a solution
of saccharin and peppermint after they had learned
a lithium-induced aversion to the taste of the
sweet-minty solution. These rats showed a reluctance to kill the "candied" mice and ate less
of them than of unflavored mice. When presented
with a candied mouse, some of the rats vigorously
pawed it as if trying to rub off the candied coat.
Some rats rubbed their chins on the floor, cower-
ed against sides of the cages and tried to climb
the walls of the cage. Rusiniak et al. considered
these responses of rats to the candied mouse to
represent displacement reactions and signs of
intense disgust. Their third experiment showed
lithium-induced aversion to the candied taste to
suppress muricide of rats on condied mice in a
T-maze. Their fourth experiment involved the use
of lithium to condition an aversion to the odor
of live candied mice in a T-maze. Administration
of lithium after rats hunted candied mice produced an aversion to the candied scent after
several days of treatment. The rats given the lithium treatment after hunting candied mice gradually learned to avoid contact with the mice on
the basis of odor cues. In general, their findings were consistent with a "two-phase" model for

predatory aggression in rats (Garcia, Rusiniak
and Brett, 1977). The first phase is the appetitive phase. It consists of plastic, coping motor
responses guided by distal cues. The second
phase is the consummatory phase. It consists of
a series of stereotypic tearing, chewing and
swallowing responses principally controlled by
taste cues. The appetitive and consummatory
phases of predatory aggression in rats are
thought to be relatively independent. Lithium-
induced aversions appear to be more readily established to consummatory behavior than to
appetitive behavior. Under certain circumstances,
however, lithium-induced learned aversions to
distal cues can occur and may influence appetitive behavior.

Smythe, Brandstater and Lazarus (1979) noted
increased fighting in groups of rats injected
with a hypertonic solution of lithium. Their
observation agrees with previous reports of
increased aggression in rats given a lithium
treatment with prompt, adverse aftereffects (see
Smith, 1977). Smythe et al. noted in addition
that an injection of 3,4-dimethoxyphenylethyl-
amine prevented fighting in rats given hypertonic lithium. Their observation suggests a
role of dopaminergic neurotransmission in the
fighting.

EFFECT OF LITHIUM ON SHOCK-INDUCED
ACTIVE AND PASSIVE AVOIDANCE IN RATS

Bignami, Pinto-Scognamiglio and Gatti (1974)
examined effects of lithium (2-9 mEq/kg) on active avoidance of rats tested in a shuttle box.
The rats had to learn to leave one side of the

shuttle box in response to a light in order to
avoid an electric footshock. In one experiment,
rats received lithium (2 mEq/kg) while learning
the active avoidance response. The lithium treatment failed to influence the acquisition of the
response by the rats. In another experiment,
lithium was administered to rats already trained
to perform the avoidance response. Under these
conditions, lithium suppressed avoidance responding soon after injection but not later on. The
time-course of lithium's effects suggested peripheral actions to be responsible for the disruption of active avoidance behavior. Bignami et
al. noted signs of intoxication in rats given
the doses of lithium with effects on active avoidance. Therefore, they considered lithium to lack
specific effects on active avoidance behavior in
rats. Berggren, Ahlenius and Engel (1980) also
studied effects of lithium (1 - 1.8 mEq/kg) on
active avoidance behavior in trained rats.
Their rats had to respond to the sound of a
buzzer in order to avoid footshock. Only the
highest dose of lithium used suppressed the conditioned avoidance response. Lithium failed to
influence escape behavior, although rats given
the highest dose appeared somewhat sedated and
displayed a moderate muscular hypotonia. Administration of L-dopa plus an inhibitor of peripheral decarboxylation counteracted the suppressant effect of lithium on conditioned avoidance
responding. Berggren et al. considered the lithium-induced suppression of conditioned avoidance behavior to be specific and not due to un-
specific, toxic effects. They ascribed the
effect of lithium mainly to interference with
catecholaminergic neurotransmission in limbic
structures in the brain.

Katz and Carroll (1977) studied effects of long-term lithium treatment (ca. 4-6 mEq/kg) on passive avoidance behavior in rats. They offered rats sweetened milk to drink followed by an electric shock. This procedure increased the latency to drink in normal rats, while lithium-treated rats showed less of an increase in their drinking latency. Katz and Carroll interpreted their findings in terms of reduction of conflict and potential anxiolytic actions of lithium. Rider and coworkers looked for changes in passive avoidance behavior in the offspring of lithium-treated rat dams (Rider, Simonson, Weng and Hsu, 1978; Hsu and Rider, 1978). The offspring were five months old at the time of tests. They learned to go down a runway to obtain water to drink. The rats received a shock via the drinking spout on one occasion. This procedure increased the latency to drink subsequently in rat offspring of normal dams, while rats exposed to lithium both prenatally and neonatally showed less of an increase in their drinking latency. The findings showed exposure to lithium during prenatal and neonatal life to interfere with passive avoidance learning later on. Rider and coworkers interpreted their findings in terms of toxic actions of lithium during early stages of development in rats. They cautioned against generalizing their conclusions to other species.

EFFECTS OF LITHIUM ON WATER INTAKE IN RATS

Recent studies confirm previous reports that short-term and long-term lithium treatments (1-5 mEq/kg) usually increase water intake in

rats (Alexander and Alexander, 1978; Christensen, Geisler, Badawi and Madsen, 1977; McCaughran
and Corcoran, 1977; Sinclair, 1974; Zakusov,
Lyubimov, Yavorskii and Fokin, 1977; Zilberman,
Kapitulnik, Feuerstein and Lichtenberg, 1979).
Administration of potassium to rats reduced the
extent of polydipsia induced by long-term lithium treatment, apparently by preventing lithium-
induced impairment of renal ability to conserve
water (Olesen and Thomsen, 1979). On the other
hand, Mailman and coworkers found administration
of lead to neonatal rats to enhance the extent
of polydipsia induced by long-term lithium treatment in adult rats (Mailman, Krigman, Mueller,
Mushak and Breese, 1978; Mailman, Breese, Krigman,
Mushak and Mueller, 1979). Mailman and coworkers
considered the effect of lead on lithium-induced
polydipsia to be mediated by central mechanisms,
a notion contested by Weeden (1979).

Recent studies found no acute effects of lithium
(0.3-5 mEq/kg) on water intake in rats given
water to drink soon after an injection of lithium
(Domjan, 1977a; Kutscher and Wright, 1977; Westbrook, Hardy and Faulks, 1979). The failure of
lithium given by injection to enhance water intake promptly in rats contrasts with previous
observations made in rats given a stomach load
of lithium (1.2 mEq/kg) (Smith and Balagura,
1972; Smith, Balagura and Lubran, 1970a). Perhaps differences in experimental conditions used
in the studies can account for the differences
observed in acute effects of lithium on water
intake in rats (Westbrook, Hardy and Faulks,
1979).

EFFECTS OF LITHIUM ON ETHANOL INTAKE
OF RATS

Nachman, Lester and Le Magnen (1970) were the
first to report effects of lithium on ethanol
intake of rats. They gave fluid-deprived rats
a solution of 6% ethanol to drink, followed by
an injection of lithium at a dose of 2.1 mEq/kg
on two occasions. The suppressant action of lithium was reduced in rats given previous experience with lithium, suggesting that the effect
of lithium on ethanol intake may be influenced
by prior experience. Ho and Tsai (1976) gave
lithium to rats with a chronic dependence on 30-
40% ethanol. Lithium, at a daily oral dose of
2.4 - 4 mEq/kg, reduced the rats' intake of
ethanol. The suppressant action of lithium on
ethanol intake under these experimental conditions may reflect toxic effects.

Several studies examined the effects of lithium
on ethanol intake in rats offered a choice between tap water and an ethanol solution to drink.
These studies measure effects of lithium on the
rats preference for ethanol. Ho and Tsai (1975)
injected lithium at a daily dose of 0.6 mEq/kg
to rats given a choice between water and 4-11%
ethanol solution to drink. The lithium treatment
reduced the rats' preference for ethanol in that
their ethanol intake decreased and water intake
increased during the treatment. Ho and Tsai
(1976) found lithium also to reduce ethanol
preference in rats selected for their relatively high preference for a 5% ethanol solution.
These rats, given lithium at a daily dose of
0.7 mEq/kg, showed a reduction in ethanol consumption and an enhancement in water intake

during the treatment period. The suppressant
effect of lithium on the preference for rats
for ethanol in both studies by Ho and Tsai stop-
ped after lithium treatment was discontinued.
They suggested that the effects of lithium on
ethanol preference might be due to influences
on the balance between serotonergic, catecholaminergic and cholinergic systems. Alexander
and Alexander (1978) also noted suppressant
effects of lithium on ethanol intake of rats.
They offered rats a choice between a 10% solution of ethanol versus water to drink. The
rats received an injection of lithium at a dose
of 0.6 mEq/kg on each of five consecutive days.
Ethanol intake decreased while water intake
increased between the third and fifth day of the
lithium treatment. Alexander and Alexander considered the most likely explanation for the
effect of lithium on ethanol intake to be related to alterations in sodium:potassium balance.
They suggested that the altered fluid intake of
the rats given lithium was a compensatory response for lithiun-induced alterations in the
internal milieu. Steinberg and McMillan (1978)
carried out two studies on effects of lithium
on rats given water and ethanol solutions for 6
hours daily. In one study, the concentration
of ethanol offered was increased from 1 to 6%
over 18 days. A lithium injection at a dose of
2 mEq/kg was given to rats during this period.
The lithium treatment suppressed the development of a preference for ethanol otherwise seen
in rats. In the other experiment, the lithium
treatment was administered to rats with an already established preference for 6% ethanol.
The lithium treatment reduced the ethanol preference of the rats. Steinberg and McMillan as

well as Alexander and Alexander noted the effect
of lithium on ethanol preference to depend on
the concentration of ethanol solution offered.
Suppressant effects of lithium on ethanol intake were more consistent in rats offered ethanol solutions at concentrations between 6% and
10% than at a lower concentration. Sinclair
(1974; 1975) carried out studies on the influences of lithium given by injection on ethanol
preference of rats. In one study, an injection
of lithium at a dose of 2 mEq/kg reduced the
intake of 6% ethanol solution without altering
water intake. In the other study, injections
of lithium at a dose of 3 mEq/kg failed to have
consistent effects on the rats' intake of 0.7%
ethanol solution versus water.

Some studies considered effects of lithium in
rats with experimentally induced preferences for
ethanol. Zakusov, Lyubimov, Yavorskii and Fokin
(1977) induced a preference for ethanol in rats
by giving them a solution of 5% ethanol to drink
as the only source of fluid for 3 hours daily
for two months. During the next two weeks,
some of the rats received an injection of lithium at a daily dose of 1.66 mEq/kg. Rats given
this lithium treatment showed a reduction in
their preference for ethanol solution. Zakusov
et al. examined also the effects of addition of
lithium to fluid on the ethanol preference of
rats. They gave rats a solution of 5% ethanol
plus lithium to drink as the only source of
fluid for several weeks. The daily lithium
dosage was probably about 2.5 mEq/kg. After
several weeks, the rats received a choice between 5% ethanol versus tap water to drink. The
lithium treatment reduced the rats' preference

for ethanol during the test. Zakusov et al.
considered the effect of lithium on ethanol
preference of rats to be mediated by actions
on neurosecretory processes of hypothalamic
centers. They described the effect of lithium
on ethanol preference in psychological terms
as a reversal of motivation. McCaughran and
Corcoran (1977) induced a chronic preference for
ethanol in rats by hypothalamic stimulation.
Injections of lithium at a daily dose of 0.6
mEq/kg reduced the rats' preference for 10-19%
ethanol solution. McCaughran and Corcoran considered the effect of lithium to depend on the
degree of preference of the rats for ethanol,
since the effects of lithium seemed to be weak-
est in rats with the most well-established ethanol preference.

Sinclair (1974; 1975) examined the effect of
addition of lithium to food on the intake of
ethanol solution by rats. One experiment involved rats given lithium to eat at a daily dose
of 0.7 - 1.07 mEq/kg. The rats showed an imme-
diate decrease in their intake of 10% ethanol
and no change in water consumption. Later on,
lithium intake increased while consumption of
ethanol remained low. The finding showed a
reduction of ethanol intake in rats given lithium in their food. Two other experiments examined the relation between the daily lithium
dosage and alterations in the rats' preference
for a 7% ethanol solution. A reduction in
ethanol intake occurred on some days during the
first week in rats given food containing lithium at a daily dose of 0.2 - 1.9 mEq/kg. Another experiment by Sinclair involved the effect
of lithium on the enhancement of ethanol intake

seen after a period of ethanol deprivation in rats with extended previous experience drinking ethanol. Lithium, added to the food only during the period of ethanol deprivation, reduced the extent to which the intake of a 7% ethanol solution was enhanced after deprivation. The lithium treatment failed to affect saccharin intake of rats, suggesting that the action of lithium may be specific for ethanol. It is unlikely that the inhibitory effect of lithium on the enhancement of ethanol intake after deprivation was due to toxic actions at the time of tests, because lithium was given at a relatively low dose and only during the deprivation period. Sinclair drew attention to the variability of lithium's effects on alcohol intake in rats and to the apparently big effects of small changes in experimental procedures.

CONDITIONED FLAVOR AVERSIONS INDUCED BY LITHIUM IN RATS

Four basic steps are involved in most studies on lithium-induced conditioned aversions in rats. First, the rats are trained to consume a substance for a short time each day. Second, they receive a novel flavor, such as that of saccharin, in the substance on one day. Third, they receive an injection of lithium soon after consuming the novel-tasting substance. Fourth, they receive the same novel-tasting substance again on a following day. The conditioned aversion is measured by the reluctance of the rats to consume the substance a second time. Rats acquire lithium-induced conditioned aversions to flavors added to foods as well as fluids (Nachman and Ashe, 1973; White, Sklar and Amit, 1977).

Rats with a lithium-induced conditioned aversion
to one flavor often show a reluctance to consume
novel substances of other flavors (Domjan, 1975a;
1977a). Toxic effects of lithium are assumed to
mediate conditioned aversions (Braveman, 1977).
Some consider conditioned aversions to provide
exceptions to general laws of learning (Seligman,
1970; Rozin and Kalat, 1971; Rozin, 1977; Garcia
and Hankins, 1977; Garcia, Rusiniak and Brett,
1977), while others include aversion conditioning in traditional learning theories (Bitterman,
1975; 1976; Revusky, 1977; Logue, 1979).

Some of the main experimental variables known to
influence the production of conditioned aversions
are the dose, time of administration and route of
administration of lithium. Increasing the dose
of lithium administered typically increases the
strength of the conditioned aversion formed
(Nachman and Ashe, 1973). Administration of
lithium soon after rats drink the novel solution
typically leads to stronger conditioned aversion
than when administration of lithium is delayed a
long time (Barker and Smith, 1974; Smith, 1978a).
An important feature of aversion conditioning
is, however, that it can occur even when administration of lithium is delayed up to several hours
after rats taste a novel solution (Garcia,
Rusiniak and Brett, 1977). Administration of
lithium by intragastric, subcutaneous and intra-
peritoneal route all produce conditioned aversions, while injection of lithium directly into
the CSF in the brain fails to induce conditioned
taste aversions in rats (Smith, 1980).

Other variables that influence the strength of
conditioned aversions are sodium deficiency, the
stimulus for thirst, the amount of the flavored

stimulus ingested, previous experience with the
flavored stimulus and previous experience with
the lithium treatment.   Smith and coworkers
showed sodium deficiency to increase the speed
with which rats overcome conditioned aversions
induced by lithium (2.4 mEq/kg) (Smith, Balagura
and Lubran, 1970b; Balagura and Smith, 1970).
Domjan (1975b) found a conditioned flavor aversion induced by lithium (0.72 -1.2 mEq/kg) to be
 more pronounced in rats drinking in response
to cellular dehydration than to water deprivation.
Domjan and Wilson (1972) investigated the role
of ingestive behavior in the production of conditioned flavor aversions.   They installed can-
nulae in the cheek of rats and administered a
saccharin solution into the mouth via the can-
nulae.   They manipulated the rate of infusion
and the degree of fluid deprivation in order to
influence the amount drunk by the rats.   The
rats received an injection of lithium (0.72 -
0.84 mEq/kg) after either tasting or tasting
and ingesting the saccharin solution. A greater
aversion to saccharin appeared in rats that
tasted and ingested saccharin than in those that
only tasted saccharin before the lithium injection.   Domjan and Wilson concluded that gustatory stimulation is sufficient for the formation
of conditioned aversions and that ingestive behaviors facilitate aversion learning.   Revusky,
Parker, Coombes and Coombes (1976) obtained
similar results.   They noted a stronger aversion
in rats given a sweet solution to drink prior to
administration of lithium (1.9 mEq/kg) than in
rats allowed only to taste the sweet solution.
Braveman and Crane (1977) investigated further
the effects of ingestive behavior on the formation of conditioned taste aversions.   In their

study, thirsty rats received up to 10.5 ml of saccharin to drink followed by a lithium injection (3 mEq/kg). An increase in the amount of saccharin consumed (up to 5.5 ml) before the lithium injection led to an increase in the strength of the conditioned taste aversion shown by the rats. This observation agreed with the findings of Domjan and Wilson (1972), so the weaker aversion seen in rats given less saccharin to drink before the lithium injection was accounted for in terms of insufficient ingestive cues. However, a further increase in the amount of saccharin consumed tended to reduce the conditioned taste aversion formed. Braveman and Crane considered either the increased amount of fluid consumed or the increased amount of time spent drinking to be responsible for the reduction in conditioned taste aversions seen in rats given the most saccharin to drink. They pointed out that the relation between the amount of a novel solution consumed before administration of lithium and the strength of the conditioned aversion formed may not always be a simple one. Best and Gembling (1977) studied effects of previous experience with a flavored solution on the formation of conditioned aversions. In the jargon of learning theory, the flavored solution is called the CS and the effect of previous experience with the solution is called the CS-preexposure effect. Best and Gembling first gave rats a solution of casein to drink on two occasions. The rats received an injection of lithium (3 mEq/kg) after drinking the casein solution for the second time. The strength of the conditioned aversion formed depended on the amount of time that elapsed between the two exposures to casein. A maximum reduction in the conditioned aversion occurred in rats given casein on two occasions at an interval of three hours. Longer intervals produced less reduction in the conditioned taste aversion. Best and Gembling considered the effect of preexposure to the taste stimulus to be due at least partly to short-term, nonassociative processes. They interpreted their findings in terms of the general information-processing model of Wagner (1976).

Previous experience with a lithium treatment tends to lessen conditioned aversions induced by lithium in rats. In the jargon of learning theories, the lithium treatment used to condition an aversion is called the "US" and the effect of previous experience with lithium is called the US-preexposure effect (Domjan and Best, 1977). In these studies, lithium is administered to rats both before and after they drink a novel solution. Cannon, Berman, Baker and Atkinson (1975) were the first to study this effect systematically. They treated rats with lithium (2.7 - 7.2 mEq/kg) at times before conditioning a flavor aversion and found the aversion formed to be reduced by the treatment in a dose-dependent manner. Revusky and coworkers also found preexposure to lithium (1.9 - 2.1 mEq/kg) to reduce the formation of a conditioned flavor aversion in rats (Revusky and Taukulis, 1975; Revusky, Parker, Coombes and Coombes, 1976). Domjan and Best (1977) examined further the US-preexposure effect. They administered lithium (1.8 mEq/kg) to rats at times before conditioning a flavor aversion. Their findings confirmed the observations of Cannon and coworkers. The US-preexposure effect appears to be a transient,

time-limited phenomenon rather than a durable,
long-lasting interference effect. Domjan and
Best considered the priming hypothesis of
Wagner (1976) and the opponent process theory of
Solomon and Corbit (1974) to provide reasonable
explanations for the effects of lithium preexposure on conditioned flavor aversions. Braveman (1977), who also observed the lithium-pre-
exposure effect, proposed that stress-related
neurohumoral changes may be involved in the
phenomenon.

DISRUPTION OF LITHIUM-INDUCED
CONDITIONED AVERSIONS IN RATS

Attempts to disrupt aversions induced by lithium
relate to interest in understanding the under-
lying mechanisms of conditioning. Several drugs
reduce lithium-induced conditioned aversions in
rats. Domjan and Wilson (1972) compared a lithium-induced aversion formed in normal rats to
that formed in rats paralyzed with d-tubocurare
to prevent swallowing. The curarized rats received a saccharin solution dropwise on their
tongue followed by an injection of lithium (2.4
mEq/kg). Less of an aversion appeared subsequently in the rats trained while under the influence of curare than in normal rats. The
findings support the notion that ingestion aids
the formation of conditioned flavor aversions.
Cappell, LeBlanc and Endrenyi (1972) found
chlordiazepoxide to disrupt a conditioned flavor
aversion induced by lithium (3 mEq/kg). They
administered chlordiazepoxide to rats with a
lithium-induced conditioned aversion to a saccharin solution and found it to enhance extinction
of the conditioned saccharin aversion. Cappell

et al. considered antianxiety effects of chlordiazepoxide as well as actions on thirst to be
involved in the disruption in the lithium-induced conditioned flavor aversion. Jolicoeur,
Wayner, Rondeau and Merkel (1978) noted a disruptive effect of phenobarbital on a conditioned
flavor aversion induced by lithium (3 mEq/kg)
in rats. Administration of phenobarbital before
the drinking test reduced the conditioned aversion. Jolicoeur et al. concluded that phenobarbital can attenuate lithium-induced taste
aversions. The disruptive effect appeared related to a general enhancement of drinking. Hennessy, Smotherman and Levine (1976) administered
dexamethasone prior to conditioning a lithium-
induced aversion in rats. The dexamethasone
treatment attenuated the conditioned flavor
aversion produced by lithium (3 mEq/kg). The
findings with dexamethasone suggest an involvement of pituitary-adrenal function in both the
acquisition and extinction of learned flavor
aversions.

A number of other studies used disruption of
conditioning to investigate neural mechanisms
for flavor aversions induced by lithium. Nachman (1970) observed only limited effects of
electroconvulsive shock treatment in the formation of lithium-induced conditioned aversion.
Rats received a saccharin solution to drink
followed by an electroconvulsive shock treatment and an injection of lithium (3 mEq/kg).
The shock treatment reduced the conditioned
flavor aversion only in rats given saccharin to
drink for a very short time (5 or 6 sec) before
application of current. The findings suggest
that neural consolidation of the flavor stimulus

occurred very rapidly in the rats. Davis and
Bures (1972) considered flavor stimuli to have
prolonged neural after-effects in rats, and
assumed neuronal information on gustatory cues
to be stored temporarily in the cerebral cortex.
They used cortical spreading depression to suppress cortical function and studied the locus of
storage of sensory information in flavor conditioning. Rats received a saccharin solution to
drink, followed by cortical spreading depression
and then an injection of lithium (2.4 mEq/kg).
Cortical spreading depression applied up to
several hours after the rats drank saccharin,
but before injection of lithium, disrupted the
formation of the conditioned flavor aversion.
The findings support the notion of a role for
cortical processes in flavor conditioning.

Danguir and Nicolaidis (1976) used paradoxical
sleep deprivation to study central mechanisms
for conditioning of lithium-induced flavor aversions. They gave rats a solution of lithium to
drink at times during paradoxical sleep deprivation, and examined the generalized conditioned
aversion formed to a saline solution. Paradoxical sleep deprivation before drinking the lithium solution reduced the generalized aversion,
while paradoxical sleep deprivation after drinking the lithium solution had little effect on
the aversion. The findings showed paradoxical
sleep deprivation to disrupt the acquisition
more than the retention of a lithium-induced
conditioned flavor aversion. Venkatakrishna-
Bhatt, Bures and Buresova (1978) carried out
further studies on effects of paradoxical sleep
deprivation on lithium-induced flavor aversions.
They induced a conditioned aversion to saccharin

in rats by an injection of lithium (3-6 mEq/kg).
They too observed a reduction in flavor aversion in rats given paradoxical sleep deprivation before conditioning. Permitting the rats
to sleep for several hours between paradoxical
sleep deprivation and conditioning reduced the
disruptive effects of paradoxical sleep deprivation on the lithium-induced aversion. The
findings support the view that paradoxical sleep
deprivation interferes with the formation of
short-term memory processes for flavors. Venkatakrishna-Bhatt et al. observed also an effect
of paradoxical sleep deprivation on the retention of the lithium-induced conditioned flavor
aversion. A reduction in the conditioned aversion to saccharin occurred in rats deprived of
paradoxical sleep only before the retention
test. Evidently, paradoxical sleep deprivation
can influence the retention of a lithium-induced
conditioned flavor aversion under certain circumstances.

Revusky and coworkers described some other ways
to disrupt the formation of a lithium-induced
aversion (Taukulis and Revusky, 1975; Revusky,
Parker and Coombes, 1977). In one experiment,
they used a rather complicated procedure called
conditioned inhibition. First, rats learned that
the presence of an odor signalled no lithium
treatment. Then, the rats received the odor
together with a novel flavor and an injection
of lithium (1.5 mEq/kg). Rats treated in this
way showed a reduction in the conditioned flavor aversion. In another experiment, they used
a procedure called overshadowing. First, rats
drank a novel solution of sugar and then a novel
solution of vinegar followed by an injection of

lithium (ca. 5.3 mEq/kg). Rats treated in this
way showed a reduction in the conditioned aversion to the sugar solution. Revusky and coworkers considered their findings to oppose the
notion that conditioning of flavor aversions
with lithium is a primitive form of learning
(Rozin and Kalat, 1971).

ENHANCEMENT OF LITHIUM-INDUCED CONDI-
TIONED TASTE AVERSION IN RATS

Lithium-induced conditioned taste aversions in
rats can be increased in several ways. One way
is to give rats a dose of either ACTH or certain
analogues of ACTH during extinction of the conditioned aversion (Hennessy, Smotherman and
Levine, 1976; Rigter and Popping, 1976). Such
a treatment delays extinction of a conditioned
aversion to a sweet solution induced in rats by
lithium (1.5 - 3 mEq/kg). Hennessy et al. considered the effect of ACTH on the lithium-induced aversion to suggest an involvement of neurohumoral feedback to the central nervous system
in conditioning, while Rigter and Popping interpreted their findings in terms of increased motivation, increased attention and increased mo-
tivational value of environmental stimuli. Two
other ways to enhance lithium-induced conditioned taste aversions are with ethanol or paradoxical sleep deprivation. Cappell, LeBlanc and
Endrenyi (1972) found administration of ethanol
before conditioning to delay extinction of a
lithium-induced aversion in rats, while Venkata-
hrishna-Bhatt, Bures and Buresova (1978) noted
enhancement of conditioned flavor aversion in
rats subjected to paradoxical sleep deprivation
before administration of lithium (6 mEq/kg).

Perhaps aversive effects of ethanol and paradoxical sleep deprivation mediated the enhancement
in lithium-induced conditioned taste aversions
in these studies. An additional way to enhance
lithium-induced conditioned flavor aversion is
to alter the novelty of stimuli. Mitchell,
Kirschbaum and Perry (1975) increased the novelty of certain stimuli by habituating rats to
other stimuli present in the environment before
conditioning. This procedure potentiated conditioning induced by lithium (3 mEq/kg). Taukulis
and Revusky (1975) made a similar observation.
They trained rats to an odor and a solution not
associated with lithium, and then presented the
rats with the same odor and another solution together with a lithium injection (1.5 mEq/kg).
The procedure increased the aversion conditioned
to the second solution. Taukulis and Revusky
considered the enhancement of conditioning to
support the applicability of general laws of
learning for lithium-induced conditioned flavor
aversions.

LITHIUM-INDUCED ENHANCEMENT OF FLUID
INTAKE IN RATS

Domjan and Gillan (1977) observed enhanced drinking of a solution offered to rats with a conditioned taste aversion to another solution.
Their experiment had two basic steps. First,
they established a lithium-induced conditioned
aversion to the taste of a particular solution
in rats. Then, they exposed rats to the taste
of that solution before offering them a second
solution to drink. Under these conditions, the
rats increased their intake of the second solution. Domjan and Gillan called the phenomenon
"intake-enhancement aftereffect of a lithium

conditioned aversive taste". They considered
three explanations to account for the phenomenon,
based on anticipatory or compensatory reactions
of rats to being poisoned.

LITHIUM-INDUCED CONDITIONED AVERSIONS TO NON-GUSTATORY STIMULI IN RATS

A distinction is to be made between studies in
which rats received taste and non-gustatory stimuli simultaneously versus studies in which these
stimuli are presented separately (Best, Best and
Mickley, 1973). Under conditions of simultaneous presentation of taste and non-gustatory stimuli, rats appear to acquire lithium-induced
aversions more readily to taste stimuli (Garcia
and Hankins, 1977). But in the absence of taste
stimuli, administration of lithium can produce
conditioned aversions to non-gustatory stimuli
in rats. Garcia, Rusiniak and Brett (1977)
pointed out that overgeneralization of conclusions reached from studies using simultaneous
presentations of taste and non-gustatory stimuli
led some to the faulty notion that rats fail to
acquire lithium-induced conditioned aversions to
non-gustatory stimuli. Revusky and Parker (1976)
used repeated doses of lithium (2 - 3.4 mEq/kg)
to establish conditioned aversions to non-gustatory stimuli in rats. They conditioned rats to
avoid either drinking water from a cup versus a
spout or drinking water from a steel spout versus
a glass spout. Administration of lithium soon
after drinking was necessary to produce conditioned aversions to non-gustatory stimuli. Their
findings show, nevertheless, that lithium can
mediate conditioned aversions in rats to the
appearance of a container. Nachman, Rauschenberger and Ashe (1977) described similar experi-

ments on the use of lithium (3 mEq/kg) to conditioned aversions to non-gustatory stimuli in rats. Lithium-induced aversions to licking a stream of air, drinking from a jar versus a spout, drinking from a wide spout versus a narrow spout, and drinking from a shallow dish versus a deep dish occurred in rats, although more than one lithium treatment was often needed for conditioning to occur. Nachman et al. suggested that either the discriminability of stimuli or the interactions of afferent neural input govern the occurrence of lithium-induced conditioned aversions to non-gustatory stimuli. Mitchell, Kirschbaum and Perry (1975) established conditioned aversions to food containers and food pellet size by two injections of lithium (3 mEq/kg) in rats. Their study showed stimulus novelty to play a role in the conditioning of non-gustatory stimuli, in that aversions occurred only in rats habituated to the environment. Evidently, the novelty of stimuli plays a role in the formation of conditioned aversions to non-gustatory stimuli just as in conditioned flavor aversions. In an unpublished study, I used lithium to condition an aversion in rats to stepping down from a platform into a black box. I gave an injection of lithium (2.5 mEq/kg) immediately after they stepped down from the platform, and left the rats in the box for the next 20 minutes. After two such treatments given three days apart, the lithium-treated rats showed a marked reluctance to step-down off the platform into the box, expressed as an increase in their step-down latency. My findings with lithium agree with those obtained in rats given apomorphine (Best et al., 1973) and show that, in the absence of oral and taste cues, lithium adminis-

tration can induce a conditoned aversion in rats
to visual-spacial stimuli. Rudy and Cheatle
(1977) used lithium to condition an aversion to
odor in neonatal rats. They placed two-day old
rat pups in a bag containing pine shavings scent-
ed with lemon and gave the pups an injection of
lithium (3 mEq/kg). Rudy and Cheatle measured
the conditioned aversion as the amount of time
the pups spent subsequently over lemon-scented
shavings in a test area. Pups given the lithium
treatment during the time of exposure to the
lemon scent in the bag tended to stay away from
the scent in the test area. Their findings are
of interest with regard both to the use of lithium to condition aversions to non-gustatory
stimuli and to the ability of an immature nervous system to support associative learning.
It is to be noted that also adult rats can acquire a lithium-induced conditioned aversion to
an odor (Taukulis and Revusky, 1975).

LITHIUM-INDUCED BACKWARD CONDITIONING
IN RATS

Three basic steps are involved in studies on
backward conditioning of a lithium-induced aversion. First, rats receive an injection of lithium. Second, they drink a novel solution. Third,
they receive the same solution to drink again on
a following day. Backward conditioning appears
as the reluctance of the rats to drink the solution. The main difference between ordinary conditioning and backward conditioning is the time
of administration of lithium. In backward conditioning, lithium is given *before* esposure to the
novel solution. In ordinary conditioning, lithium is given *after* exposure to the novel solu-

tion.   Interest in backward conditioning stems
mainly from the fact that traditional learning
theories considered it to occur rarely if at all.

Boland (1973) failed to observe lithium-induced
backward conditioning in rats, but Barker and
Smith (1974) described backward conditioning of
an aversion in rats given lithium (3 mEq/kg).
Later, they pointed out that backward conditioning appears to be weaker than conditioning obtained by ordinary procedures (Barker, Smith
and Suarez, 1977).   In addition, they suggested
that backward conditioning may be ordinary conditioning mediated by a long delay in the onset of
lithium's toxic effects.   Domjan and coworkers
carried out a series of experiments on lithium-
induced backward conditioning (Domjan and Best,
1977; Domjan and Gregg, 1977; Gillan and Domjan,
1977).   They used lithium at a dose of 1.8 -2.25
mEq/kg and found the strength of lithium-induced
backward conditioning to be influenced by the
time of administration of lithium, the concentration of the novel solution drunk after the lithium injection, and exposure to another novel
solution before the lithium injection.   The
strength of backward conditioning probably also
depends on lithium dosage.   Their findings are
consistent with the notion that aversions produced by backward conditioningare weaker than
those formed by ordinary conditioning.   Nevertheless, they oppose the notions that backward
conditioning is due to ordinary conditioning and
that taste-aversion learning is a primitive form
of conditioning.   The failure to observe backward conditioning in some studies may reflect
the ineffectiveness of the conditioning proce-

dures used in those studies rather than the brev-
ity of lithium's effects.

ILLNESS-MEDIATED NEOPHOBIA IN RATS
GIVEN LITHIUM

Two basic steps are involved in studies on illness-mediated neophobia in rats (Domjan, 1977a).
First, rats receive an injection of lithium.
Second, they are offered a novel solution to
drink. Illness-mediated neophobia appears as a
reluctance of the lithium-treated rats to drink
the novel solution. Barker and Smith (1974)
noted illness-mediated neophobia in thirsty rats
given lithium (3 mEq/kg) shorthy before drinking
a sweet solution. Domjan (1975; 1977a; 1977b)
carried out systematic studies on this type of
neophobia in rats. Administration of lithium
(0.3 -3 mEq/kg) reduced the subsequent consumption of novel solutions by rats in a dose-dependent manner. Lithium-induced neophobia was short-
lasting and corresponded closely to the time
during which the rats looked sick. It occurred
in response to every novel solution tested
(casein, vinegar and saccharin). The novelty of
the solution was an important variable for the
production of illness-mediated neophobia. Rats
raised on either a sweet solution or a sour solution failed to reduce their intake of the fa-
miliar solution after an injection of lithium.
Illness-mediated neophobia occurred, however,
despite short-term familiarization of rats with
a solution before administration of lithium.
Domjan considered lithium-mediated neophobia induced by lithium to be a sensitivation effect of
poisoningrather than a special case of taste-
aversion conditioning. Evidently, rats are

hyperreactive to novel flavors offered soon after administration of lithium. Kutscher and Wright (1977) also observed lithium-induced neophobia in rats given lithium 2.2- 3.3 mEq/kg). They found lithium to suppress subsequent intake of a bitter solution. Short-term familiarization to the bitter solution failed to prevent the suppressant effects of lithium on fluid intake of the rats, in accordance with Domjan's findings. Illness-mediated neophobia induced by lithium appears to be an unconditioned aversion related to a reduction in the acceptability of novel solutions.

BEHAVIOR OF "LITHIUM-DEFICIENT" RATS

Patt, Pickett and O'Dell (1978) carried out a unique study on effects of too little lithium in rats. They fed up to three generations of rats a diet with a very low lithium content. The daily lithium intake of rats given the low-lithium diet was 50-150 nEq/kg, while the daily lithium dose for those on the normal diet was 3-5 μEq/kg. They observed no behavioral abnormalities in the rats on the low-lithium diet and no reliable alterations in fertility and growth. Their evidence fails to establish lithium as an essential trace element with an important physiological role in rats.

BEHAVIOR OF LITHIUM-INTOXICATED RATS

The main symptoms of lithium intoxication in rats are lethargy, ataxia, pronounced inactivity, prostration, reduced food intake, tremor, irritability and body weight loss. These symptoms appear regularly in rats given an injection of lithium at a dose of 3 mEq/kg or more (Nachman, Lester and Le Magnen, 1970; Barker and Smith, 1974; Danguir, Nicolaidis and Perino-Martel, 1976; Braveman, 1977; Barker, Smith and Suarez, 1977). Some symptoms of lithium intoxication occur occasionally in rats given a single lithium injection at a dose between 2 and 3 mEq/kg (Etevenon, Guillon, Breteau and Boissier, 1971; Domjan and Wilson, 1972; Davis and Bures, 1972; Kutscher and Wright, 1977; Domjan and Gillan, 1977; Smith 1978a; Johnson and Jonson, 1978; Smythe, Brandstater and Lazarus, 1979), while repeated administration of lithium at a daily dose of 2-5 mEq/kg led regularly to symptoms of intoxication in rats (Holmes, Rodnight and Kapoor, 1977; Samples and Seybold, 1977). The adverse effects of repeated lithium injections in this dose range were less in rats provided with a salt-lick or kept from becoming wet due to the increased amount of urine usually excreted by rats given lithium (Hesketh, Kinloch and Reading, 1977; Plenge, 1978). On the other hand, lithium toxicity was greater in rats given concurrent treatment with physostigmine or certain neuroleptic drugs (Beck and Reis, 1976; Samples, Janowsky, Pechnick and Judd, 1977). Daily injections of lithium at a dose greater than 6 mEq/kg usually caused death in rats, often after a few days with muscle twitching and hyperreactivity (Jacobs, 1978; Thomsen and Olesen, 1978).

Chapter 7

EFFECTS OF LITHIUM ON GUINEA PIGS, HAMSTERS AND RABBITS

GENERAL EFFECTS OF LITHIUM ON BEHAVIOR OF GUINEA PIGS

Recent studies provided some information on dose-response effects of lithium on behavior of guinea pigs. Balfour, Hewick and Murray (1979) noted no apparent signs of sedation in guinea pigs given lithium (1-4 mEq/kg) for 7 days. Hirsch, Ehrenpreis and Comaty (1978) gave guinea pigs long-term treatment with lithium (0.3 - 5 mEq/kg). The treatment appeared to be nontoxic, in that the animals continued to gain weight steadily. Cade (1978) commented further on his pioneering studies on effects of lithium on behavior in guinea pigs. Administration of lithium (ca.5-6 mEq/kg) caused guinea pigs to be lethargic and unresponsive. When turned onto their backs, guinea pigs given lithium lay there and gazed placidly instead of showing usual frantic righting behavior. Taken together, these studies show a dose of lithium below ca. 5 mEq /kg to be without marked overt effects on behavior of guinea pigs.

CONDITIONED AVERSIONS IN GUINEA PIGS GIVEN LITHIUM

Braveman (1974) used lithium (3 mEq/kg) to induce conditioned aversions in guinea pigs. In one experiment, the guinea pigs received lithium after drinking a dilute solution of saccharin or of hydrochloric acid. Subsequently, the guinea pigs showed a reluctance to drink the solution previously paired with lithium treatment. In another experiment, administration of lithium conditioned an aversion to a colored solution. It is noteworthy that guinea pigs lack cones in their retina, so the conditioned aversion probably was based on the brightness of the solution rather than on its color. Extinction of the lithium-induced conditioned aversion was faster for the color discrimination than for the taste discrimination. Nevertheless, the findings show guinea pigs to acquire conditioned aversions to either visual or gustatory cues related to fluid consumption. Evidently, a lithium dose of 3 mEq/kg has strong enough effects in guinea pigs to mediate conditioned aversions.

EFFECT OF LITHIUM ON DRUG-INDUCED STEREOTYPIC BEHAVIOR IN GUINEA PIGS

Klawans and coworkers examined effects of long-term lithium treatment on gnawing induced by *d*-amphetamine or apomorphine in guinea pigs (Klawans, Hitri, Nausieda and Weiner, 1977; Klawans, Weiner and Nausieda, 1977). The guinea pigs re-

ceived a diet containing lithium, but the daily
lithium dose consumed is uncertain. Administration of lithium together with haloperidol prevented the potentiation of drug-induced gnawing
otherwise seen in guinea pigs subjected to the
haloperidol pretreatment, while administration
of lithium after haloperidol failed to affect
the drug-induced stereotypy. Klawans and coworkers interpreted their findings in terms of
lithium-induced prevention of the development
of hypersensitiviy of dopamine receptors.

EFFECTS OF LITHIUM ON BEHAVIOR
OF HAMSTERS

Three studies investigated aspects of behavior
of hamsters (*Mesocricetus auratus*) given lithium.
Hofmann, Günderoth-Palmowski, Wiedenmann and
Engelmann (1978) reported preliminary data on
the circadian rhythm of hamsters given lithium
(1.2 - 1.8 mEq/kg) in their drinking water. The
hamsters lived under conditions of continuous
red lighting. A slight increase in the period
of the circadian rhythm occurred in some of the
hamsters during lithium treatment. Johnston
and Zahorik (1975) used lithium (2 mEq/kg) to
produce a conditioned aversion in hamsters.
Their studies involved a pheromone-like substance, namely the sexual attractant in the
vaginal secretion of female hamsters. Male
hamsters given lithium after licking the vaginal
secretion from a glass plate were reluctant to
approach and to lick a glass plate coated with
the secretion again subsequently. The findings
show lithium to alter the response of male hamsters to the substance and indicate that the
response of animals to pheromones is modifiable.

Opitz (1977) used hamsters to investigate effects of lithium (5 mEq/kg) on ethanol intake. It is noteworthy that hamsters normally show a propensity for solutions of ethanol. Hamsters offered an ethanol solution versus water drank much more of the ethanol solution than of the water. Short-term administration of lithium reduced the preference of the guinea pigs for the ethanol solution. Opitz emphasized the importance of selection of the animal species best suited for studies on lithium's behavioral actions.

EFFECTS OF LITHIUM ON SEIZURES IN RABBITS

Boissier (1958) found a high dose of lithium (11.4 mEq/kg) to reduce the phase of clonic seizure induced by electroshock in rabbits, while lower doses of lithium (2.9 - 5.7 mEq/kg) failed to affect shock-induced seizures. On the other hand, Ammar and Osman (1977) found lithium (0.12 mEq/kg) to cause clonic and tonic seizures in rabbits. They administered lithium directly into the cerebroventricles. Rabbits given lithium in this way showed protrusion of eyeballs, erection of the tail, intermittent spasm of neck muscles and seizures. All the rabbits given lithium died as a result of the treatment, apparently from respiratory failure.

Chapter 8

EFFECTS OF LITHIUM ON FELINES AND FERRETS

GENERAL EFFECTS OF LITHIUM ON BEHAVIOR IN CATS

Administration of lithium (ca. 2.1 μEq/kg) directly into the cerebrospinal fluid of awake cats failed to affect their behavior, in that none of the cats showed tremor, seizures or mydriasis (Garcy and Marotta, 1978). Low daily doses of lithium (0.6 mEq/kg) in sustained release tablets failed to alter the behavior of cats, except for producing mild lethargy (Marini, Williams and Sheard, 1978). A single moderate dose of lithium (0.8 -1.4 mEq/kg) increased total sleep time in cats during a short observation period (Zhukov, 1977). Lithium also increased sleep in cats after they received stress ful brain stimulation. Moderate daily doses of lithium (0.8 -1.4 mEq/kg) usually caused cats to decrease their intake of food and water, to lose weight, and to show fine generalized tremor

(Marini, Williams and Sheard, 1978; Lanoir, 1979). High daily doses of lithium (2-4 mEq/kg) produced signs of intoxication such as vomiting, diarrhea, increased mewing while awake, hyper-excitability to stimulation and tremor, but no spontaneous seizures (Lanoir and Lardennois, 1977; Lanoir, 1979). Hypothalamic stimulation provoked convulsions in one cat given long-term treatment with lithium (0.8 mEq/kg) (Marini, Williams and Sheard, 1978) while an intravenous injection of lithium (2.4 mEq/kg) appeared to reduce convulsant effects of pentylenetetrazole in cats (Lanoir and Lardennois, 1977).

EFFECT OF LITHIUM ON AGGRESSION IN CATS

Sheard (1977) mentioned inhibitory effects of lithium on attack behavior elicited by electri-cal stimulation in the brain of cats. The cats received lithium orally at an unspecified dose. The lithium treatment reduced attack behavior elicited by stimulation of hypothalamic sites, without affecting other elicited behaviors such as hissing and growling. Marini, Williams and Sheard (1978) gave cats daily doses of lithium (0.3 -2.3 mEq/kg) in slow-release tablets for up to 2 weeks. They failed to note a change in the well-directed attack of cats at the ex-perimentor or in response to electrical stimu-lation in the brain.

LITHIUM-INDUCED CONDITIONED AVERSION IN A COUGAR

Garcia and coworkers used lithium to condition an aversion to venison in a captive cougar (*Felis concolor*) named Fred (Gustavson, Kelly,

Sweeney and Garcia, 1976; Garcia, Rusiniak and Brett, 1977).   Its normal diet consisted of cow and horse meat.   Fred received lithium (ca. 3.2 mEq/kg) as coated tablets in a portion of ground venison on one occasion.   Subsequently, it approached a dish containing venison, turned over the dish, examined and mouth each chunk of venison, but refrained from eating the meat.   Evidently, the reluctance of the cougar to eat the venison was a consequence of the lithium treatment.   Garcia et al. considered the behavior of the cougar toward the venison to suggest that the lithium treatment altered the hedonic tone of the meat.

CONDITIONED AVERSIONS INDUCED BY LITHIUM IN FERRETS

Ferrets (*Mustela putorius*) are carnivores related to mink and weasels.   They normally prey on rodents, small rabbits and birds.   Rusiniak, Gustavson, Hankins and Garcia (1976) carried out studies on lithium-induced conditioned aversions in captive, tame ferrets.   In one experiment, thirsty ferrets received a solution of lithium to drink on several occasions.   The amount of lithium consumed exceeded 2.7 mEq/kg. The ferrets usually vomited soon after drinking the lithium solution.   Only a weak conditioned aversion appeared to the lithium solution as well as to another salt solution.   Another experiment investigated conditioned aversions to food.   Ferrets received either dog food or fish to eat in place of their normal mink food ration.   Then, they received lithium (1.8 mEq/kg) by injection.   A strong conditioned aversion to either dog food or fish occurred subsequently in the ferrets given lithium. The third experiment considered effects of lithium on mouse-killing behavior. Ferrets typically kill their prey by biting at the back of the neck. Hungry ferrets promptly killed mice. Rusiniak et al. offered ferrets mice to kill, but not to eat. The ferrets received an oral dose of lithium (37.8 mEq/kg) soon after killing mice on several occasions. The lithium treatment induced copious vomiting, but failed to inhibit subsequent mouse-killing by the ferrets. Injections of lithium (4.5 mEq/kg) also failed to prevent mouse-killing. Rusiniak et al. noted, however, some signs of conditioned aversions to mice in ferrets treated with lithium. When confronted with a mouse, the lithium-treated ferrets often retched and groomed and made no oral contact with the prey. Instead of killing the mouse with a bite at the neck, the lithium-treated ferrets swiftly trampled the mice to death! Rusiniak et al. interpreted their findings in terms of a "two-phase" notion of predatory behavior. They considered lithium to inhibit the consummatory phase but not the appetitive phase of predation in ferrets.

Chapter 9

EFFECTS OF LITHIUM ON CANINES AND BEARS

GENERAL EFFECTS OF LITHIUM ON BEHAVIOR
IN DOGS

McDonald, Schemehorn and Stookey (1978) administered lithium to dogs in their drinking
water at a concentration of 0.6 - 2.4 mEq/kg
for up to four weeks. The daily lithium dose
consumed is uncertain. Some of the dogs refused
to eat and lost weight after two weeks of lithium treatment, and one dog died during treatment.
The cause of death appeared to be acute renal
failure.

EFFECTS OF LITHIUM ON AGGRESSION IN DOGS

Hornstein (1975) noted increased aggression in
a dog given daily oral doses of lithium (ca. 4.1
mEq/kg) for 25 days. After 15 days of treatment,
the dog became aggressive, restive and difficult
to handle. It continued to be aggressive until
the end of the experiment. In contrast, Horn-

stein and coworkers observed a reduction in aggression in another experiment on a dog given
lithium (Hornstein, de Alencar Filho, de Toledo
and Spirck, 1977). Prior to lithium treatment,
the dog had to be gagged for experiments because
it was aggressive and hard to handle. The dog
received electric current via cotton swabs soaked with a solution of lithium and applied in an
orbital mastoideal position. The treatment induced electrosleep. Afterwards, the dog showed
no aggression and failed to bark for three weeks.
Thereafter, it continued to lack aggression,
although it reacted to stimuli and barked normally during 18 months of observation.

EFFECTS OF LITHIUM ON PREDATORY BEHAVIOR
IN CAPTIVE COYOTES

The effect of lithium on predation by coyotes
(*Canis latrans*) is controversial. Some studies
show lithium treatment to reduce the natural
tendency of coyotes to attack, kill and eat prey,
while other studies find lithium treatment to be
without reliable effects on coyote predation.
The controversy began with a study by Gustavson,
Garcia, Hawkins and Rusiniak (1974) in which
lithium was given to captive coyotes in association with consumption of certain prey (see Smith,
1977). Gustavson et al. considered their findings to show that food aversions, established
by lithium administration in captive coyotes,
inhibited attack behavior in certain living prey.
In a subsequent study, Gustavson, Kelly, Sweeney
and Garcia (1976) used lithium-adulterated food
to induce food aversions in captive coyotes.
Six coyotes were used. They were deprived of
food for 48 hours before tests. Three of the

coyotes, named Josh, Draba and Pacer, learned
to go down a 25 m long runway to obtain a tethered rabbit and a tethered chicken. These
coyotes promptly attacked and killed both prey,
and took them back to their home cages to eat.
After four such tests, Josh and Draba received
a lithium-adulterated bait which consisted of
71 mEq LiCl in dog food sewn into a fresh rabbit hide. The lithium dose consumed by these
coyotes was about 11.6 mEq/kg body weight. The
coyotes vomited about 90 minutes after eating
this bait. They each were offered a rabbit in
the next test. They each attacked and killed it,
but did not eat it. Josh urinated on his rabbit. The lithium-adulterated bait offered to
Pacer was dead rabbit injected with 143 mEq
LiCl. Pacer vomited about 30 minutes after eating his bait. A live rabbit as well as a live
chicken were offered in the next test. Pacer
attacked, killed and ate the chicken but left
the rabbit unharmed. In the next test carried
out two days later, Pacer attacked and killed
the rabbit and left the chicken unharmed. The
test situation was modified for the three other
coyotes. One of them, named Girl, received dog
food instead of chicken to eat. The two other
coyotes, named Annie and Meg, failed to perform
in the test area so rabbit and chicken were
given to them in their home cages. After eating
a lithium-adulterated rabbit bait, Girl did not
attack a live rabbit but ate dog food readily.
Meg and Annie ate a lithium-adulterated rabbit
carcass and, in the next test, ate a lithium-
adulterated dog food bait wrapped in rabbit
flesh. in subsequent tests, Annie killed and
ate chickens but not rabbits, while Meg killed
and ate chickens and killed rabbits, but did not

eat them. Gustavson and coworkers concluded
that lithium-adulterated baits can be used to
induce aversions in coyotes to certain live prey.

Conover, Francik and Miller (1977) were unable
to induce reliable food aversions to live prey
in captive coyotes given lithium together with
a dead prey. Their first experiment involved
five coyotes that all normally ate dead and live
chickens. They received dead chickens laced
with 95-143 mEq lithium and all coyotes consumed
at least one entire lithium-laced chicken. The
lithium dose consumed was about 3.8 -7.2 mEq/kg
body weight. It produced emesis in the coyotes.
Subsequently, the coyotes showed a reluctance to
eat dead chickens, but they all continued to
kill and eat live chickens. The second experiment gave similar results. Four coyotes received 1-3 injections of lithium at a dose of ca.
2.4 mEq/kg after eating dead mice. Emesis occurred 2-4 hours after the lithium injection. The
coyotes were offered either dead mice, live mice
or both dead and live mice during subsequent
tests. An aversion to dead mice occurred, but
the coyotes continued to kill and eat live mice
readily. The aversion to dead mice began to
decline after nine days of testing. Nevertheless, each coyote ate far fewer dead mice than
live mice overall during 24 days of testing.
Conover et al. concluded that their coyotes failed to become averted to live prey species after
being subjected to aversive conditioning with
dead prey. Griffiths, Conolly, Burns and Sterner (1978) also were unable to prevent prey-kil-
ling in captive coyotes given lithium. In one
study, coyotes became averted to chicken carcasses injected with 143-167 mEq lithium, but con-

tinued to kill and to eat live chickens. In
another study, four experienced sheep-killer
coyotes and one naive coyote received ground
lamb meat or sheep carcass containing lithium
for three days. By the third day, each coyote
appeared to eat carefully, rejecting salty-
tasting portions of the carcass. The coyotes
subsequently killed live lambs at the first
opportunity. Griffiths et al. note that the
negative findings in some studies and positive
findings in others may be due to differences in
the details of experimental design, such as bait
material and prey, lithium dosage, route of lithium administration, cirteria for aversion and
the length of test trials.

FIELD STUDIES ON EFFECTS OF LITHIUM
ON PREDATORY BEHAVIOR IN COYOTES

Gustavson et al. (1976) carried out a field
study on the use of lithium-adulterated baits
against coyote predation on sheep. The study
took place on 3000 acres of the Honn Ranch in
the state of Washington. Twelve bait sessions
were established on the sheep-grazing area.
The baits were made of dog food and 143 mEq LiCl
wrapped and stapled in fresh, unsheared sheep
hide. In addition to these baits, carcasses of
sheep found dead during the study were sprayed
with a solution of 1.96 M LiCl and added to the
bait stations. A record was kept of the number
of sheep and lambs killed by coyotes during the
study. The criteria of coyote predation was
evidence of puncture wounds in the neck, dry
blood on the wool, and indications of hemorrhag-
ing on dead sheep. The presence of snow during
the first part of the study facilitated tracking

the coyotes. The number of sheep killed by coyotes during the study was compared to the rancher's records for sheep losses during the same
period for the three previous years. The number
of sheep killed by coyotes tappered off and
plateaued after about 50 days. After 72 days,
the rancher and the researchers began to disa-
gree on the number of sheep killed by coyotes,
with more kills judged as new kills by the rancher than by the researchers. The study continued
for about one more month. According to the
rancher, coyote predation was reduced by 30%
during the study with lithium, while the researchers' records showed a reduction in coyote
predation of 60%. Gustavson et al. mentioned
that increased human activity in the area, in
addition to the presence of lithium baits, may
have driven coyotes from the area and thereby
reduced predation. Support for this notion came
from the observation that no coyotes were sited
in the area during the last month of the study.
While the data suggest that coyote predation of
sheep was reduced during the study compared to
previous years, the reasons for the reduction
are uncertain.

Ellins, Catalano and Schechinger (1977) carried
out a field study on coyote predation in Southern
California. They too measured the number of
sheep and lambs killed by coyotes. One herd of
sheep and lambs grazed in an area bordered by
10 bait stations containing dead sheep and lambs
injected with a solution of 0.94 M LiCl, while
another herd grazed in an area bordered by bait
stations with dead sheep and lambs injected with
a sodium cloride solution. The two herds were
about 12.5 km apart. Information on the number

of sheep killed by coyotes was obtained 2-3 times per week form sheep herders for 10-18 weeks. Coyotes killed about 14 animals per week in the herd with lithium baits during the first 7 weeks, and 1-2 per week during the next 11 weeks of the study. The number of sheep and lambs killed by coyotes was about 19 per week for the first 2 weeks, and then leveled off to 1-2 per week during the next 8 weeks in the herd with sodium chloride baits. Ellins et al. considered their findings to show the taste aversion method of lithium-baiting to be useful in controlling coyote predation.

Griffiths, Conolly, Burns and Sterner (1978) reexamined field studies on the use of lithium to prevent coyote predation. They questioned the notion that lithium had reliable deterent effects on sheep and lamb killings by coyotes. They reanalysed some previous data in terms of the total number of animals killed as well as in terms of the number of animals killed per week of the field studies. They found no effect of lithium on coyote predation in results expressed as kills per week. Griffiths et al. pointed out that coyotes were subject to trapping, shooting and strychnine baits in addition to the lithium baits in most previous studies. Griffiths et al. also mentioned a field study carried out by members of the Saskatchewan Department of Agriculture in which sheep meat baits containing lithium were placed on farms and community sheep pastures. Information on the number of sheep killed by coyotes during the two years of the study was obtained at the end of each year and was compared to estimates for the number of sheep lost to coyotes in the pre-

vious year. A reduction in sheep losses to coyotes occurred during the years of lithium bait-
ing. However, the absence of information on
coyote predation in comparable areas without
lithium baits during the years of the study and
the use of supplementary measures against coyote predation, such as traps, snares and shooting coyotes in addition to confining small lambs,
removing carrion and applying lethal poison other
than lithium to fresh kills precludes definite
conclusions on the role of lithium in the reduction observed in coyote predation during the
study. Griffiths et al. concluded that more
tightly designed field studies are needed to
determine whether lithium is of use in the prevention of coyote predation. They outlined
some guidelines for such studies.

CONDITIONED AVERSION INDUCED BY LITHIUM IN WOLVES

Garcia and coworkers used lithium to induce a
conditioned aversion to sheep in two captive
wolves (*Canis lupus*) (Gustavson, Kelly, Sweeney
and Garcia, 1976; Garcia, Rusiniak and Brett,
1977). Before lithium treatment, the wolves
immediately attached sheep, seized them in
their jaws, pulled them down and killed them
with prolonged biting. Garcia et al. gave the
wolves a sheep hide containing sheep flesh and
lithium (ca. 5.1 mEq/kg) to eat on one occasion.
Subsequently, the wolves charged a sheep and
made oral contact with it by characteristic
flank attacks, but then immediately released it.
The wolves withdrew from the sheep and respond-
ed submissively to it during the test. When
offered rabbits, the wolves immediately attacked,

killed, and ate them and left the sheep unharmed. Evidently, the wolves acquired a conditioned aversion to sheep after one experience with
lithium-adulterated mutton. Whether lithium
baits can avert wolves to sheep in the wild is
an open question. Garcia et al. interpreted
their findings in terms of lithium-induced alterations in the hedonic value of taste stimuli.

CONDITIONED AVERSIONS IN BEARS GIVEN LITHIUM

Gustavson (1977) mentioned studies on the use
of lithium to condition aversions in American
black bears (*Ursus americanus*). The bears received lithium (0.8 -2.6 mEq/kg) to eat in a
marshmellow and/or by injection soon after eating a marshmellow. The lithium treatment usually caused the bears to vomit. Subsequently,
most of the bears rejected marshmellows offered
to eat from time-to-time over several weeks.

Chapter 10

EFFECTS OF LITHIUM ON MONKEYS

EFFECT OF LITHIUM ON AGGRESSION IN MONKEYS

Brocco, Slikker and Killan (1976) administered lithium to an adolescent monkey (*Macacca mulatto*). It received lithium at a daily dose of 0.2 mEq/kg for two weeks and a daily dose of 0.4 mEq/kg for the next 4 days. The monkey showed a transient loss of aggressivity and reactivity to people during the third week of treatment. Discontinuation of lithium treatment failed to influence the behavior of the monkey.

CONDITIONED AVERSION IN MONKEYS GIVEN LITHIUM

Johnson, Beaton and Hall (1975) used lithium to establish conditioned aversions to visual stimuli in monkeys (*Cercopithecus sabaeus*). The monkeys received an injection of lithium

(0.76 or 1.5 mEq/kg) after drinking water from a yellow or blue bottle. Subsequently, the monkeys refrained from drinking from a bottle of the same color previously associated with administration of lithium at a dose of 1.5 mEq /kg, while the lower lithium dose failed to produce a conditioned aversion. The findings show administration of lithium to produce a conditioned aversion based on visual stimuli in monkeys. In addition, they suggest a lithium dose of 0.75 mEq/kg to be without pronounced adverse aftereffects in the species of monkey used.

Chapter 11

CRITICAL COMMENTS

   Only few advances were made in recent years
toward a better understanding of lithium's
influence on animal behavior. I consider the
work on lithium-induced conditioned aversions
to be the most informative. The results obtained in those studies permit conclusions to be
drawn on the influence of experimental variables
such as lithium dosage, time of lithium treatment, and previous experience with lithium on
the formation of conditioned flavor aversions
in rats. In addition, the findings show lithium-
induced aversions to occur throughout most of
the animal kingdom and probably to be mediated
by toxic actions of lithium outside the central
nervous system (CNS). The main unanswered
question of studies on lithium-induced conditioned aversions is the nature of the unconditioned response. Braveman (1977) suggested the
unconditioned response to be a physiological
change related to the "stress" of the lithium
treatment. Fluctuations in the level of adrenal

gland hormones can be ruled out as an essential
unconditioned response, because lithium-induced
conditioned aversions occur in rats lacking
adrenal glands (Smith, Balagura and Lubran,
1970b). Further research is needed to determine
whether effects on other stress-related physiological systems serve as the unconditioned response
in the production of lithium-induced conditioned
aversions.

Many studies on effects of lithium on animal
behavior did not deal with conditioned aversions
but were rather aimed at throwing light on factors related to the psychiatric uses of lithium.
Most of these studies were, however, too incom-
plete to permit conclusive statements on actions
of lithium. As a rule, too little attention
was given to the species selected, the lithium
treatment used, and the behavioral test employed.
Consequently, several basic questions remain unanswered about the effects of lithium on behavior
on laboratory animals. One question is the
extent to which behavioral actions of lithium
are due to central effects. My working definition of a central effect is an action in the
CNS as opposed to a peripheral effect which is
an action outside the CNS. Prompt behavioral
changes in animals given lithium by injection
may be mediated primarily by peripheral effects
since the injection usually leads to high concentrations of lithium in the blood stream and
extracerebral tissues while the concentration of
lithium in the CNS rises very slowly. One way
to study primarily central effects of lithium
is to administer lithium directly in the brain.
Studies carried out on several species show central injections of lithium to influence only

some aspects of behavior (Stern, Lekovic, Przic
and Casparovic, 1961; Watts and Mark, 1970;
Mark and Watts, 1971; Benowitz and Sperry, 1973;
Smialowski, 1976; Inoue, Tsukada and Barbeau,
1977; Smith, 1980). Studies on behavior of animals given central as well as peripheral injections of lithium are needed to permit the rela-
tive role of central and peripheral actions of
lithium on behavior to be determined.

A second question is the extent to which actions
of lithium on animal behavior are due to specific
effects. My working definition of a specific
effect is an action unique to lithium, as opposed
to a nonspecific effect which is an action
common to lithium and some other substances.
Specific effects of lithium on animal behavior
are of interest in order to clarify whether
lithium may have unique actions on some psychiatric disorders. One way to search for specific
actions of lithium is to compare behavioral
effects of lithium to several other substances
in animals in the same test situation (e.g.,
Sanger and Steinberg, 1974; Smith, 1978a). This
approach is relatively rare, however, so little
is known with certainty about the extent to
which actions of lithium on animal behavior are
specific.

A third question is the extent to which a particular behavioral effect of lithium is a general action. My working definition of a general
action is an effect produced in a variety of
species, as opposed to a solitary action observed
in one or only a few species. General actions
of lithium in laboratory animals are of interest
in order to find effects possibly related to

actions of lithium in humans. Only when a particular effect of lithium is observed in a wide
variety of species can we expect that effect
also to occur in humans. On the other hand,
solitary effects of lithium may also be of use
to clarify the behavioral actions of lithium.
It is of interest, for example, to know the
causes of differences on the influence of lithium on alcohol preference in rats, hamsters and
mice (Sinclair, 1974; 1975; Ho and Tsai, 1975;
1976; Opitz, 1977; Nottage, Syme and Syme, 1978;
Alexander and Alexander, 1978; Steinberg and
McMillan, 1978). An understanding of the causes
of these differences may provide information on
possible causes of the differences in effects
of lithium treatment on alcoholism in humans
(Schou, 1979b; Kline and Cooper, 1979; Merry
and Coppen, 1979).

A fourth question is the extent to which actions
of lithium on animal behavior are primary effects.
My working definition of a primary effect is an
action directly on the target organ as opposed to
a secondary effect caused by an indirect action.
Interest in distinguishing between primary and
secondary effects stems from the desire to know
the chain of causal events of lithium's behavioral actions. It is unsettled, for example,
whether increased water intake of animals given
lithium treatments reflects mainly a primary
effect on central regulation of thirst or a
primary effect on kidney function and a secondary action on thirst (see Smith, 1977; Westbrook,
Hardy and Faulks, 1979). One way to distinguish
between primary and secondary actions of lithium
may be to note the temporal sequence of effects.
Another way may be to remove target organs one-
by-one and to note which actions of lithium

remain and which ones are eliminated. A know-
ledge of the causal chain-of-events for lithium's
actions on behavior of animals may provide clues
as to primary causes of some therapeutic actions
and side effects of lithium in humans. In my
opinion, further studies are needed on the chain-
of-causation of lithium's actions on animal behavior.

The fifth question is the extent to which actions
of lithium on animal behavior are due to toxic
effects. My working definition of a toxic effect
is an action which, if continued, would cause
premature death, as opposed to a nontoxic effect
which would not cause premature death. In general, more attention was given to the question
of lithium toxicity in recent studies on animal
behavior than in the past. Many recent studies
administered lithium in ways that failed to
produce overt signs of distress in the animals.
Administration of lithium either together with
adequate amounts of sodium and potassium in the
food or as twice daily injections of low doses
in dilute solutions appears to minimize lithium
toxicity. The use of these and similar procedures of lithium administration in the future
may provide information on nontoxic actions on
animal behavior.

The final point to be considered is probably a
minor one. It concerns the preparation of a
lithium solution from the chloride salt. The
hygroscopic nature of this salt may make preparation of a solution with a particular lithium concentration problematic, although many
may be unaware of the difficulty. Jernigan,
Schrank and Kraus (1978) noted, however, a discrepancy of about 9% between the expected and the

actual concentration of lithium in a solution
made from lithium chloride. I too usually observe a discrepancy between the expected and
the actual concentration of lithium in solutions
made directly from reagent grade lithium chloride
salt stored in a dessicator. I consider it
necessary to determine the concentration of lithium in solution analytically in order to be
sure of it. Many may have overlooked this detail,
however, so the actual lithium dose administered
in some studies probably was less than was in-
tended. It may be advisable in the future to
determine the actual concentration of lithium
in solutions used in studies on animal behavior.

Chapter 12

SUMMARY OF MAJOR FINDINGS

1. Addition of lithium to sea water influenced the swimming rhythm of jellyfish.

2. The rate of movement of marine amoeba depended on the ration between lithium and calcium in solution.

3. The circadian rhythm of cockroaches in running wheels increased after addition of lithium to their drinking water.

4. An injection of lithium induced aversions to food in codfish.

5. Goldfish tended to swim together less after several hours of exposure to lithium in solution.

6. Exposure to lithium in solution for several days influenced choice-point behavior of goldfish in a maze.

7. Hawks given a lithium injection learned to avoid eating bitter-tasting mice.

8. Young chicks learned a conditioned aversion after one lithium treatment.

9. Pigeons showed increased water intake after an injection of lithium.

10. Lithium failed to affect spontaneous activity reliably in mice.

11. Lithium restored curiosity-related activity of isolated mice to normal levels.

12. Lithium suppressed some behaviors and potentiated others induced by amphetamine in mice.

13. Head-twitching occurred in mice given lithium together with reserpine, tetrabenazine, syrosingopine, or 5-methoxy-N, N-dimethyl-tryptamine.

14. Lithium and no reliable effects on ethanol intake in mice.

15. Mice given lithium solution to drink showed reduced susceptibility to seizures induced by nicotine.

16. Lithium administration potentiated ethanol-narcosis in mice.

17. Injections of lithium inhibited isolation-induced aggression in mice.

18. Administration of lithium to mice after they showed predatory aggression inhibited this behavior subsequently.

19. Daily doses of lithium failed to affect maternal aggression in mice.

20. Toxic actions of lithium in mice depended on genetic factors, route of administration, concurrent drug treatment and time of administration, but not on the anion used.

21. No prompt effects of lithium appeared on locomotor activity of rats tested in automatic activity recorders.

22. Lithium treatment increased exploration of novel stimuli by rats.

23. Daily injections of lithium potentiated behaviors induced by fenfluramine, 5-hydroxytryptophan and 5-methoxy-N, N-dimethyltryptamine in rats.

24. Lithium treatment counteracted running wheel activity produced by *l*-amphetamine, methamphetamine and apomorphine in rats.

25. Daily adminsitration of lithium counteracted hyperactivity induced by thyroid hormone treatment in young rats.

26. Stereotypic behavior induced by phenylethylamine or apomorphine was suppressed by daily doses of lithium.

27. Injections of lithium potentiated head-

twitches induced by 5-methoxy-N, N-dimethyl-
tryptamine, 5-methoxytryptophan and 5-
methoxytryptamine in rats.

28. Catalepsy induced by reserpine or neuroleptic drugs in rats was potentiated by high
doses of lithium.

29. Lithium enhanced seizures induced by several
experimental treatments in rats.

30. Paradoxical sleep was reduced promptly
in rats given high doses of lithium.

31. Rats usually showed a reduction in aggressive behavior after lithium treatment.

32. A high dose of lithium suppressed shock-induced active avoidance behavior in rats.

33. Long-term lithium treatment disrupted
passive avoidance behavior in rats.

34. Water intake usually increased in rats
given short-term or long-term lithium treatments.

35. Daily doses of lithium usually reduced the
preference rats showed for an ethanol
solution.

36. A single injection of lithium produced
reliable conditioned aversions to novel
flavored substances in rats.

37. The strength of lithium-induced conditioned
taste aversions was influenced by previous
experience in rats.

38. Administration of chlordiazepoxide, pheno-
    barbital or dexamethasone disrupted condi-
    tioned aversions induced by lithium in rats.

39. Lithium-induced conditioned taste aversions
    in rats were enhanced by ACTH, ethanol and
    stimulus novelty.

40. Administration of lithium mediated condi-
    tioned aversions to several types of non-
    gustatory stimuli in rats.

41. Backward conditioning to a novel taste
    stimulus occurred in rats given lithium.

42. Rats were reluctant to drink a novel solu-
    tion offered soon after administration of
    a single lithium injection.

43. Administration of a diet containing very
    little lithium failed to have untoward
    effects in rats.

44. Doses of lithium above 2 mEq/kg typically
    led to overt toxic effects in rats.

45. Loss of righting reflex occurred in
    guinea pigs given lithium.

46. A conditioned aversion to a visual stimulus
    appeared in guinea pigs given lithium.

47. Administration of lithium together with
    haloperidol prevented potentiation of stereo-
    typic behavior induced in guinea pigs given
    amphetamine or apomorphine.

48. A conditioned aversion to a pheromone occur-
    red in hamsters given lithium.

49. Hamsters given daily lithium doses showed a reduction in their propensity for an ethanol solution.

50. Moderate doses of lithium caused cats to show reduced food intake, generalized tremor and increased total sleep time.

51. Lithium administration induced a conditioned aversion to venison in a cougar.

52. Ferrets given lithium in association with mouse-killing refrained from eating mice, but continued to kill them.

53. Lithium treatments induced conditioned aversions to some foods offered to captive coyotes.

54. Predatory behavior of captive coyotes failed to be prevented reliably by lithium treatments.

55. Inconclusive findigns were obtained in field studies using lithium to control predatory behavior of coyotes.

56. An aversion to live bait appeared in captive wolves given a lithium treatment.

57. Bears usually acquired a lithium-induced conditioned aversion to marshmellows.

58. An injection of lithium induced a conditioned aversion to a visual stimulus in monkeys.

ACKNOWLEDGEMENTS

I thank the Danish Medical Research Council and P. Carl Petersens Fund for financial support and Mogens Schou for constructive criticism.

REFERENCES

Abdallah AH, Riley CC, Boeckler WH, White JA.
The effect of subchronic administration of
lithium carbonate on the pharmacological activity of d-amphetamine. Pharmocologist 19:
227, 1977.

Alexander GJ, Alexander RB. Alcohol consumption
in rats treated with lithium carbonate or
rubidium chloride. Pharmacol Biochem Behav
8: 533-6, 1978.

Allikmets LH, Stanley M, Gershon S. The effect
of lithium on chronic haloperidol enhanced
apomorphine aggression in rats. Life Sci 25:
165-70, 1979.

Ammar EM, Osman FH. Electrocardiographic and
neurological changes following injection of
monovalent and divalent cations into the
cerebral ventricle of conscious rabbits.
Pharmacol Res Commun 9: 563-72, 1977.

Arnsten AT, Segal DS. Naloxone alters Locomotion
and interaction with environmental stimuli.
Life Sci 25: 1035-42, 1979.

Balagura S, Smith DF. Role of LiCl and environmental stimuli on generalized learned aversion
to NaCl in the rat. Am J Physiol 219: 1231-4,
1970.

Balfour D, Hewick D, Murray N. Comparison of plasma, erythrocyte and brain lithium concentrations in the guinea-pig and rat. Br. J Pharmacol 67; 474P-5P, 1979.

Banister EW, Bhakthan NMG, Singh AK. Lithium protection against oxygen toxicity in rats: Ammonia and amino acid metabolism. J Physiol 260: 587-96, 1976.

Barker LM, Smith JC. A camparison of taste aversions induced by radiation and lithium chloride in CS-US and US-CS paradigms. J Comp Physiol Psychol 87: 644-54, 1974.

Barker LM, Smith JC, Suarez EM. "Sickness" and the backward conditioning of taste aversions. pp 533-53 in Learning Mechanisms in Food Selection, ed LM Barker, MR Best, M Domjan. Baylor University Press, Waco Texas, 1977.

Beck PJ, Reis DJ. A toxic interaction between lithium and some psychotropic agents in rat. Res Commun Psychol Psychiat Behav 1: 269-82, 1976.

Benowitz LI, Sperry RW. Amnesic effects of lithium chloride in chicks. Exp Neurol 40: 540-6, 1973.

Berg D, Baenninger R. Predation: Separation of aggressive and hunger motivation by conditioned aversion. J. Comp Physiol Psychol 86: 601-6, 1974.

Berggren U, Ahlenius S, Engel J. Effects of acute lithium administration on conditioned avoidance behavior and monoamine synthesis in rats. J Neural Transmission 47: 1-10, 1980.

Berggren U. Tallstedt L, Ahlenius S, Engel J. The effect of lithium on amphetamine-induced locomotor stimulation. Psychopharmacology 59: 41-5, 1978.

Best MR, Gemberling GA. Role of short-term processes in the conditioned stimulus preexposure effect and the delay of reinforcement gradient in long-delay taste-aversion learning. J Exp Psychol: Anim Behav Processes 3: 253-63, 1977.

Best PJ, Best MR, Mickley GA. Conditioned aversion to distinct environmental stimuli resulting from gastrointestinal distress. J Comp Physiol Psychol 85: 250-7, 1973.

Bignami G, Pinto-Scognamiglio W, Gatti GL. The evaluation of the behavioural toxicity of psychotropic agents: The case of lithium. Proc Euro Soc Study Drug Toxic 15: 33-42, 1974.

Bitterman ME. The comparative analysis of learning. Are the laws of learning the same in all animals? Science 188: 699-709, 1975.

Bitterman ME. Flavor aversion studies. Science 192: 266-7, 1976.

Boissier J-R. Actions pharmacologiques du lithium, sa place parmi less métaux alcalins. Actualites Pharm 11: 5-40, 1958.

Boland FJ. Saccharin aversion induced by lithium chloride toxicosis in a backward conditioning paradigm. Animal Learn Behav 1: 3-4, 1973.

Borison RL, Sabelli HC, Maple PJ, Havdala HS, Diamond BI. Lithium prevention of amphetamine-induced 'manic' excitement and of reserpine-induced 'depression' in mice: Possible role of 2-phenylethylamine. Psychopharmacology 59: 259-62, 1978.

Brain PF, Al-Maliki S. Effects of lithium chloride injections on rank-related fighting, maternal aggression and locust-killing responses in naive and experienced 'TO' strain mice. Pharmacol Biochem Behav 10: 663-9, 1979.

Branchey MH, Cavazos LA, Cooper TB. Effects of lithium on seizure susceptibility in alcoholized and non-alcoholized rats. Commun Psychopharmacol 1: 213-24, 1977.

Braveman NS. Poison-based avoidance learning with flavored or colored water in guinea pigs. Learn Motiv 5: 182-94, 1974.

Braveman NS. What studies on preexposure to pharmacological agents tell us about the nature of the aversion-inducing agent? pp 511-30 in Learning Mechanisms in Food Selection, ed LM Barker, MR Best, M Domjan. Baylor University Press, Waco Texas, 1977.

Braveman NS, Crane J. Amount consumed and the formation of conditioned taste aversions. Behav Biol 21: 470-7, 1977.

Brett LP, Hankins WG, Garcia J. Prey-lithium aversions. III: Buteo hawks. Behav Biol 17: 87-98, 1976.

Britton DR, Bianchine JR, Greenberg S. Lithium-induced suppression of locomotor activity in rats. Pharmacologist 18: 197, 1976.

Brocco MJ, Slikker W Jr, Killam KF Jr. Characterization of EEG effects following acute and chronic administration of lithium chloride in *Macacca Mulatta*. Proc West Pharmacol Soc 19: 428-31, 1976.

Bulaev VM, Ostrovskaya RU. Effect of cesium, lithium and rubidium on some actions of morphine. Byull Eksp Biol Med 86: 42-4, 1978.

Cade JFJ. Lithium - Past, present and future. pp 5-16 in Lithium in Medical Practice, ed FN Johnson, S Johnson. MTP Press, Lancaster, 1978.

Cannon DS, Berman RF, Baker TB, Atkinson CA. Effect of preconditioning unconditioned stimulus experience in learned taste aversions. J. Exp Psychol: Anim Behav Processes 104: 270-84, 1975.

Cappell H, LeBlanc AE, Endrenyi L. Effects of chlordiazepoxide and ethanol on the extinction

of a conditioned taste aversion. Physiol
Behav 9: 167-9, 1972.

Christensen S, Geisler A, Badawi I, Madsen SN.
Plasma and urinary cyclic AMP levels in normal
and lithium-treated rats. Letters to Editor.
Acta Pharmacol Toxicol (Kbh) 40: 455-9, 1977.

Conover MR, Francik JG, Miller DE. An experimental evaluation of aversion conditioning for
controlling coyote predation. J Wildl Manage
41: 775-9, 1977.

Danguir J. Nicolaidis S. Impairments of learned
aversion acquisition following paradoxical
sleep deprivation in the rat. Physiol Behav
17: 489-92, 1976.

Danguir J, Nicolaidis S, Perino-Martel M-C.
Effects of lithium chloride on sleep patterns
in the rat. Pharmacol Biochem Behav 5: 547-
50, 1976.

Davis JL, Bures J. Disruption of saccharin-aver-
sion learning in rats by cortical spreading
depression in the CS-US interval. J Comp
Physiol Psychol 80: 398-402, 1972.

Dessaigne S, Scotto AM, Guigues M. Modifications
par le lithium des propriétés cataleptigènes
de quelques neuroleptiques chez le Rat. CR
Soc Biol 172: 1173-80, 1978.

Domjan M. Poison-induced neophobia in rats:
Role of stimulus generalization of conditioned
taste aversions. Animal Learn Behav 3: 205-
11, 1975a.

Domjan M. The nature of the thirst stimulus:
A factor in conditioned taste-aversion behavior.
Physiol Behav 14: 809-13, 1975b.

Domjan M. Selective suppression of drinking
during a limited period following aversive
drug treatment in rats. J Exp Psychol: Anim
Behav Processes 3: 66-76, 1977a.

Domjan M. Attenuation and enhancement of neophobia for edible substances. pp 151- 79 in Learning Mechanisms in Food Selection, ed LM Barker, MR Best, M Domjan. Baylor University Press, Waco Texas, 1977b.

Domjan M. Best MR. Paradoxical effects of proximal unconditioned stimulus preexposure. Interference with and conditioning of a taste aversion. J Exp Psychol: Anim Behav Processes 3: 310-21, 1977.

Domjan M, Gillan DJ. Aftereffects of lithium-conditioned stimuli on consummatory behavior. J Exp Psychol: Anim Behav Processes 3: 322-34, 1977.

Domjan M, Gregg B. Long-delay backward taste aversion conditioning with lithium. Physiol Behav 18: 59-62, 1977.

Domjan M, Wilson NE. Contribution of ingestive behaviors to taste-aversion learning in the rat. J Comp Physiol Psychol 80: 403-12, 1972.

Downie SE, Wasnidge C, Floto F, Robinson GA. Lithium-induced inhibition of ^{125}I accumulation by thyroids and growing oocytes of Japanese Quail. Poultry Sci 56: 1254-8, 1977.

Eichelman B, Seagraves E, Barchas J. Alkali metal cations: Effects on isolation-induced aggression in the mouse. Pharmacol Biochem Behav 7: 407-9, 1977.

Ellins SR, Catalano SM, Schechinger SA: Conditioned taste aversion: A field application to coyote predation on sheep. Behav Biol 20: 91-5, 1977.

Etevenon P, Fraisse B, Guillon G, Breteau G, Boissier JR. EEG et lithium cérébral chez le rat. pp 15-26 in Advances in Neuro-Psychopharmacology, ed O Vinár, Z Votava, PB Bradley. North-Holland, Amsterdam, 1971.

Flemenbaum A. Lithium inhibition of norepinephrine and dopamine receptors. Biol Psychiat 12: 563-72, 1977a.

Flemenbaum A. Antagonism of behavioral effects of cocaine by lithium. Pharmacol Biochem Behav 7: 83-5, 1977b.

Friedman E, Dallob A, Levine G. The effect of long-term lithium treatment on reserpine-induced supersensitivity in dopaminergic and serotoninergic transmission. Life Sci 25: 1263-6, 1979.

Furukawa T, Yamada K, Kohno Y, Nagasaki N. Brain serotonin metabolism with relation to the head twitches elicited by lithium in combination with reserpine in mice. Pharmacol Biochem Behav 10: 547-9, 1979.

Garcia J. Hankins WG. On the origin of food aversion paradigms. pp 3-19 in Learning Mechanisms in Food Selection, ed LM Barker, MR Best, M Domjan. Baylor University Press, Waco Texas 1977.

Garcia J, Rusiniak KW, Brett LP. Conditioning food-illness aversions in wild animals: *Caveant Cononici*. pp 273-316 in Operant-Pavlovian Interactions, ed H Davis, HMB Hurwitz. Lawrence Erlbaum Associates: Hillsdale New Jersey 1977.

Garcy AM, Marotta SF. Effects of cerebroventricular perfusion with monovalent and divalent cations on plasma cortisol of conscious cats. Neuroendocrinology 26: 32-40, 1978.

Gaston KE. An illness-induced conditioned aversion in domestic chicks: one-trial learning with a long delay of reinforcement. Behav Biol 20: 441-53, 1977.

Gillan DJ, Domjan M. Taste aversion conditioning with expected versus unexpected drug treatment.

J. Exp Psychol: Anim Behav Processes 3: 297-309, 1977.

Greist JH, Jefferson JW, Combs AM, Schou M, Thomas A. The lithium librarian. Arch Gen Psychiat 34: 456-9, 1977.

Griffiths RE Jr, Connolly GE, Burns RJ, Sterner RT. Coyotes, sheep and lithium chloride. pp 190-6 in Proc 8th Vertebrate Pest Control Conf, Univ Calif, Davis, 1978.

Gustavson CR. Comparative and field aspects of learned food aversions. pp 23-43 in Learning Mechanisms in Food Selection, ed LM Barker, MR Best, M Domjan. Baylor University Press, Waco Texas, 1977.

Gustavson CR, Garcia J, Hankins WG, Rusiniak KW. Coyote predation control by aversive conditioning. Science 184: 581-3, 1974.

Gustavson CR, Kelly DJ, Sweeney M, Garcia J. Prey-lithium aversions. 1. Coyotes and wolves. Behav Biol 17: 61-72, 1976.

Harrison-Read PE. Models of lithium action based on behavioural studies using animals. pp 289-303 in Lithium in Medical Practice, ed FN Johnson, S Johnson. MTP Press, Lancaster, 1978.

Harrison-Read PE. Evidence from behavioral reactions to fenfluramine, 5-hydroxytryptophan, and 5-methoxy-N,N-dimethyltryptamine for differential effects of short-term and long-term lithium on indoleaminergic mechanisms in rats. Br J Pharmacol 66: 144P-5P, 1979.

Hawkins R, Kripke DF, Janowsky DS. Circadian rhythm of lithium toxicity in mice. Psychopharmacology 56: 113-4, 1978.

Hennessy JW, Smotherman WP, Levine S. Conditioned taste aversion and the pituitary-adrenal system. Behav Biol 16: 413-24, 1976.

Hesketh JE, Kinloch N, Reading HW. The effects of

lithium on ATPase activity in subcellular
fractions from rat brain. J Neurochem 29:
883-94, 1977.

Hirsch J, Ehrenpreis S, Comaty JE. Effect of lithium chloride on electrically stimulated
guinea-pig longitudinal muscle-myenteric plexus.
Arch Int Pharmacodyn 232: 4-13, 1978.

Ho AKS, Ho CC. Potentiation of lithium toxicity
by ethanol in rats and mice. Alcoholism: Clin
Exp Res 2: 386-91, 1978.

Ho ARK, Ho CC. Toxic interactions with other
central depressants: Antagonism by naloxone to
narcosis and lethality. Pharmacol Biochem
Behav 11: 111-4, 1979.

Ho AKS, Tsai CS. Lithium and ethanol preference.
J Pharm Pharmacol 27: 58-9, 1975.

Ho AKS, Tsai CS. Effects of lithium on alcohol
preference and withdrawal. Ann NY Acad Sci
273: 371-7, 1976.

Hoffmann C, Smith DF. Lithium and rubidium:
Effects on the rhythmic swimming movement of
jellyfish (*Aurelia aurita*). Experientia 35:
1177-8, 1979.

Hofmann K, Günderoth-Palmowski M, Wiedenmann G,
Engelmann W. Further evidence for period
lengthening effect of Li^+ on circadian rhythms.
Z Naturforsch 33c: 231-4, 1978.

Holmes H, Rodnight R, Kapoor R. Effects of electroshock and drugs administered *in vivo* on
protein kinase activity in rat brain. Pharmacol
Biochem Behav 6: 415-9, 1977.

Hornstein SR. Lithium. Indication of selective
distribution in central nervous system. A
preliminary report. J Bras Psiquiat 24: 343-
8, 1975.

Hornstein DR, de Alencar Filho RA, de Toledo M,
Spirck CN. Preliminary data on a new method

for lithium therapy. Rev Brasil Pespuisas Med Biol 10: 249-54, 1977.

Hsu JM, Rider AA. Effect of maternal lithium ingestion on biochemical and behavioural characteristics of rat pups. pp 279-87 in Lithium in Medical Practice, ed FN Johnson, S Johnson. MTP Press, Lancaster, 1978.

Inoue N, Tsukada Y, Barbeau A. Effects of manganese, magnesium and lithium on the ouabain-induced seizure. Fol Psychiat Neurol Jpn 31: 645-51, 1977.

Jacobs JJ. Effect of lithium chloride on adreno-cortical function in the rat. Proc Soc Exp Biol Med 157: 163-7, 1978.

Janowsky DS, Abrams AA, Groom GP, Judd LL, Cloptin P. Lithium administration antagonizes cholinergic behavioral effects in rats. Psychopharmacology 63: 147-50, 1979.

Jernigan HM Jr, Schrank GD, Kraus LM. Lithium chloride and dilithium carbamyl phosphate: Lithium distribution and toxicity in mice. ToxicolAppl Pharmacol 44: 413-21, 1978.

Johnson C, Beaton R, Hall K. Poison-based avoidance learning in nonhuman primates: Use of visual cues. Physiol Behav 14: 403-7, 1975.

Johnson FN. The variety of models proposed for the therapeutic actions of lithium. pp 305-27 in Lithium in Medical Practice, ed FN Johnson, S Johnson. MTP Press, Lancaster, 1978.

Johnson FN. Lithium effects on social aggregation in the goldfish (*Carassius Auratus*). Med Biol 57: 102-6, 1979a.

Johnson FN. Effects of lithium on visual perceptual thresholds in the goldfish (*Carassius Auratus*). Neurosci Lett 11: 111-4, 1979b.

Johnson FN. The effects of lithium chloride on response to salient and nonsallient stimuli in *Carassius Auratus*. Int J Neurosci 9: 185-90, 1979c.

Johnson FN. The psychopharmacology of lithium. Neurosci Bio-behav Rev 3: 15-30, 1979d.

Johnson FN. The effects of lithium chloride on spontaneous alternation behaviour in the goldfish (*Carassius auratus*). Neuropsycho-biology 6: 72-8, 1980.

Johnson FN, Johnson S. Effects of intraperito-neal injections of hypertonic lithium cloride solutions in rats. IRCS Med Sci 6: 308-9, 1978.

Johnston RE, Zahorik DM. Taste aversions to sexual attractants. Science 189: 893-4, 1975.

Jolicoeur FB, Wayner MJ, Rondeau DB, Merkel AD. The effects of phenobarbital on lithium chloride induced taste aversion. Pharmacol Biochem Behav 9: 845-7, 1978.

Kadzielawa K. Inhibition of the action of anti-convulsants by lithium treatment. Pharmacol Biochem Behav 10: 917-21, 1979.

Katz RJ, Carroll BJ. Effects of chronic lithium and rubidium administration upon experimental-ly induced conflict behavior. Progr Neuro-Psychopharmacol 1: 285-8, 1977.

Klawans HL, Hitri A, Nausieda PA, Weiner WJ. Animal models of dyskinesia. pp 351-63 in Animal Models in Psychiatry and Neurology, ed I Hanin, E Usdin. Pergamon Press, Oxford, 1977.

Klawans HL, Weiner WJ, Nausieda PA. The effect of lithium on an animal model of tardive dyskinesia. Progr Neuro-Psychopharmacol 1: 53-60, 1977.

Kline NS, Cooper TB. Lithium in the treatment of chronic alcoholism. pp 122-8 in Lithium. Con-troversies and Unresolved Issues, ed TB Cooper,

S Gershon, NS Kline, M Schou. Excerpta Medica, Amsterdam, 1979.

Klunder CS, O'Boyle M. Suppression of predatory behaviors in laboratory mice following lithium chloride injections or electric shock. Animal Learn Behav 7: 13-6, 1979.

Kutscher CL, Wright WA. Unconditioned taste aversion to quinine induced by injections of NaCl and LiCl: Dissociation of aversion from cellular dehydration. Physiol Behav 18: 87-94, 1977.

Lanoir J. Lithium and states of alertness. pp 157-67 in **Pharmacology of States of Alertness**, ed Passonant, Oswald. Pergamon Press, Oxford, 1979.

Lanoir J, Lardennois D. The action of lithium carbonate on the sleep-waking cycle in the cat. EEG Clin Neurophysiol 42: 676-99, 1977.

Lazarus JH, Muston HL. The effect of lithium on the iodide concentrating mechanism in mouse salivary gland. Acta Pharmacol Txicol (Kbh) 43: 55-8, 1978.

Lieberman KW, Alexander GJ, Stokes P. Dissimilar effects of lithium isotopes on motility in rats. Pharmacol Biochem Behav 10: 933-5, 1979.

Logue AW. Taste aversion and the generality of the laws of learning. Psychol Bull 86: 276-96, 1979.

Lowe WC, O'Boyle M. Suppression of cricket killing and eating in laboratory mice following lithium chloride injections. Physiol Behav 17; 427-30, 1976.

McCaughran JA Jr, Corcoran ME. Lithium reduces preference for ethanol induced by hypothalamic stimulation. J Pharmac Pharmacol 29: 120-1, 1977.

McDonald JL Jr, Schemehorn BR, Stookey GK. Effect

of lithium upon plaque and gingivitis in the Beagle Dog. J Dent Res 57: 474, 1978.

MacKay B. Conditioned food aversion produced by toxicosis in atlantic cod. Behav Biol 12: 347-55, 1974.

Mailman RB, Breese GR, Krigman MR, Mushak P, Mueller RA. Lead enhancement of lithium-induced polydipsia. Reply to Weeden. Science 205: 726, 1979.

Mailman RB, Krigman MR, Mueller RA, Mushak P, Breese GR. Lead exposure during infancy permanently increases lithium-induced polydipsia. Science 201: 637-9, 1978.

Malick JB. Inhibition of fighting in isolated mice following repeated administration of lithium chloride. Pharmacol Biochem Behav 8: 579-81, 1978.

Marini JL, Sheard MH, Kosten T. Study on the role of serotonin in lithium action using shock-elicited fighting. Commun Psychopharmacol 3: 225-33, 1979.

Marini JL, Williams SP, Sheard MH. Repeated sustained-release lithium carbonate administration to cats. Toxicol Appl Pharmacol 43: 559-67, 1978.

Mark RF, Watts ME. Drug inhibition of memory formation in chickens. Proc R Soc Lond (Biol) 178: 439-54, 1971.

Martin GM, Bellingham WP, Storlien LH. Effects of varied color experience on chickens' formation of color and texture aversions. Physiol Behav 18: 415-20, 1977.

Maté C, Ribas B, Acobettro RI, Santos Ruiz A. Lithium and rubidium effects on the motor activity of rats. An Real Acad Farm 45: 279-86, 1979.

Merry J, Coppen A. Two-year follow-up of alcoholic patients formerly treated with lithium/placebo.

pp 129-33 in Lithium. Controversies and Unresolved Issues, ed TB Cooper, S Gershon, NS Kline, M Schou. Excerpta Medica, Amsterdam, 1979.

Messiha FS. Alkali metal ions and ethanol narcosis in mice. Pharmacology 14: 153-7, 1976.

Mitchell D, Kirschbaum EH, Perry RL. Effects of neophobia and habituation on the poison-induced avoidance of exteroceptive stimuli in the rat. J Exp Psychol: Anim Behav Processes 104: 47-55, 1975.

Mukherjee BP, Bailey PT, Pradhan SN. Correlation of lithium effects on motor activity with its brain concentration in rats. Neuropharmacology 16: 241-4, 1977.

Mukherjee BP, Pradhan SN. Effects of lithium on septal hyperexcitability and muricidal behavior in rats. Res Commun Psychol Psychiat Behav 1: 241-7, 1976a.

Mukherjee BP, Pradhan SN. Effects of lithium on foot shock-induced aggressive behavior in rats. Arch Int Pharmacodyn 222: 125-31, 1976b.

Murphy DL. Animal models for mania. pp 211-23 in Animal Models in Psychiatry and Neurology, ed I Hanin, E Usdin. Pergamon Press, Oxford, 1977.

Nachman M. Limited effects of electroconvulsive shock on memory of taste stimulation. J Comp Physiol Psychol 73: 31-7, 1970.

Nachman M, Ashe JH. Learned taste aversions in rats as a function of dosage, concentration, and route of administration of LiCl. Physiol Behav 10: 73-8, 1973.

Nachman M, Lester D, Le Magnen J. Alcohol aversion in the rat: Behavioral assessment of noxious drug effects. Science 168: 1244-6, 1970.

Nachman M, Rauschenberger J, Ashe JH. Studies of learned aversions using non-gustatory stimuli. pp 395-417 in <u>Learning Mechanisms in Food Selection</u>, ed LM Barker, MR. Best, M Domjan. Baylor University Press, Waco Texas, 1977.

Nottage RM, Syme GJ, Syme LA. Effect of lithium on voluntary alcohol consumption by low and high preference mouse strains. Med Biol 56: 28-31, 1978.

Olesen OV, Thomsen K. Potassium prevention of lithium-induced loss of water and sodium in rats. Toxicol Appl Pharmacol 51: 497-502, 1979.

Opitz K. Volitional ethanol intake and ethanol preference in hamsters: Drug-induced alterations. IRCS Med Sci 5: 468, 1977.

Pantin CFA. On the physiology of amoeboid movement. III. The action of calcium. Br J Exp Biol 3: 275-96, 1926.

Patt EL, Pickett EE, O'Dell BL. Effect of dietary lithium levels on tissue lithium concentrations, growth rate, and reproduction in the rat. Bioinorg Chem 9: 299-310, 1978.

Perez-Cruet J, Dancey JT. Thymus gland involution induced by lithium chloride. Experientia 33: 646-8, 1977.

Pert A, Rosenblatt JE, Sivit C, Pert CB, Bunney WE Jr. Long-term treatment with lithium prevents the development of dopamine receptor supersensitivity. Science 201: 171-3, 1978.

Plenge P. Lithium effects on rat brain glucose metabolism in long-term lithium-treated rats studied *in vivo*. Psychopharmacology 58: 317-22, 1978.

Poirel C, Hengartner O, Briand M. Circadian rhythms and periodicity analyses of emotional behaviours in mice treated with lithium.

pp 171-8 in Chronopharmacology, Proc Satellite Symp 7th Int Congr Pharmacol, Paris July 21-24, 1978, ed A Reinberg, F Halberg. Pergamon Press, Oxford, 1979.

Rastogi RB, Singhal RL. Lithium: Modification of behavioral activity and brain biogenic amines in developing hyperthyroid rats. J Pharmacol Exp Ther 201: 92-102, 1977a.

Rastogi RB, Singhal RL. Lithium suppresses elevated behavioural activity and brain catecholamines in developing hyperthyroid rats. Can J Physiol Pharmacol 55: 490-5, 1977b.

Revusky SH. Learning as a general process with an emphasis on data from feeding experiments. pp 182-97 in Food Aversion Learning, ed NW Milgram, L Krames, TM Alloway. Plenum Press, New York, 1977.

Revusky S, Parker LA. Aversions to unflavored water and cup drinking produced by delayed sickness. J Exp Psychol: Anim Behav Processes 2: 342-53, 1976.

Revusky, S, Parker LA, Coombes J, Coombes S. Rat data which suggest alcoholic beverages should be swallowed during chemical aversion therapy, not just tasted. Behav Res Ther 14: 189-94, 1976.

Revusky S, Parker LA, Coombes S. Flavor aversion learning: Extinction of the aversion to an interfering flavor after conditioning does not affect the aversion to the reference flavor. Behav Biol 19: 503-8, 1977.

Revusky S, Taukulis H. Effects of alcohol and lithium habituation on the development of alcohol aversions through contingent lithium injection. Behav Res Ther 13: 163-6, 1975.

Rider AA, Simonson M, Weng YS, Hsu JM. Effect on rat pup growth and behavior of maternal lithium

ingestion and low protein diet. Nutr Rep Int
17: 595-606, 1978.

Rigter H, Popping A. Hormonal influences on the
extinction of conditioned taste aversion.
Psychopharmacologia 46: 255-61, 1976.

Rozin P. The significance of learning mechanisms
in food selection: Some biology, psychology and
sociology of science. pp 577-89 in Learning
Mechanisms in Food Selection, ed LM Barker,
MR Best, M Domjan. Baylor University Press,
Waco Texas, 1977.

Rozin P, Kalat JW. Specific hungers and poison
avoidance as adaptive specializations of learning. Psychol Rev 78: 459-86, 1971.

Rudy JW, Cheatle MD. Odor-aversion learning in
neonatal rats. Science 198: 845-6, 1977.

Rusiniak KW, Gustavson CR, Hankins WG, Garcia J.
Prey-lithium aversions. II: Laboratory rats
and ferrets. Behav Biol 17: 73-85, 1976.

Samples JR, Janowsky DS, Pechnick R, Judd LL.
Lethal effects of physostigmine plus lithium
in rats. Psychopharmacology 52: 307-9, 1977.

Samples JR, Seybold ME. The electrophysiological
effects of lithium in the rat. Int Pharmacopsy-
chiatry 12: 160-5, 1977.

Sanger DJ, Steinberg H. Inhibition of scopolamine-
induced stimulation of Y-maze activity by
methyl-p-tyrosine and by lithium. Eur J Pharmacol 28: 344-9, 1974.

Schou M. Bibliography on the biology and pharmacology of lithium. 5. Neuropsychobiology 4:
40-64, 1978.

Schou M. Bibliography on the biology and pharmacology of lithium. 6. Neuropsychobiology 5:
241-65, 1979a.

Schou M. Alcoholism: Introduction. pp 120-1 in
Lithium. Controversies and Unresolved Issues,

ed TB Cooper, S Gershon, NS Kline, M Schou.
Excerpta Medica, Amsterdam, 1979b.

Schou M. Bibliography on the biology and pharmacology of lithium. 7. Neuropsychobiology 6:
1-28, 1980.

Seligman MEP. On the generality of the laws of
learning. Psychol Rev 77: 406-18, 1970.

Sheard MH. Animal models of aggressive behavior.
pp 247-57 in Animal Models in Psychiatry and
Neurology, ed I Hanin, E Usdin. Pergamon Press,
Oxford, 1977.

Sinclair JD. Lithium-induced suppression of
alcohol drinking by rats. Med Biol 52: 133-6,
1974.

Sinclair JD. The effects of lithium on voluntary
alcohol consumption by rats. The effects of
centrally active drugs on voluntary alcohol
consumption. Finn Found Alcohol Stud 24: 119-
42, 1975.

Smialowski A. Influence of lithium chloride on
rabbit's EEG and behavior. Pol J Pharmacol
Pharm 28: 181-7, 1976.

Smith DF. Lithium and animal behavior. In Annual
Research Reviews, vol 1, ed DF Horrobin. Eden
Press, Montreal, 1977.

Smith DF. Learned aversion and rearing movement
in rats given LiCl, $PbCl_2$ or NaCl. Experientia
34: 1200-1, 1978a.

Smith DF. Lithium chloride toxicity and pharma-
codynamics in inbred mice. Acta Pharmacol
Toxicol (Kbh) 43: 51-4, 1978b.

Smith DF. The effects of lithium on phenylethylamine behavior in rats are counteracted by
monoamine oxidase A and B inhibitors. Arch
Int Pharmacodyn 233: 221-6, 1978c.

Smith DF. Central and peripheral effects of lithium on conditioned taste aversions in rats.

Psychopharmacology 1980 (in Press).

Smith DF, Balagura S. Sodium appetite in rats given lithium. Life Sci 1021-9, 1972.

Smith DF, Balagura S, Lubran M. "Antidotal thirst": A response to intoxication. Science 167: 297-8, 1970a.

Smith DF, Balagura S, Lubran M. Some effects of adrenalectomy on LiCl intake and excretion in the rat. Am J Physiol 218: 751-4, 1970b.

Smythe GA, Brandstater JF, Lazarus L. Acute effects of lithium on central dopamine and serotonin activity reflected by inhibition of prolactin and growth hormone secretion in the rat. Aust J Biol Sci 32: 329-34, 1979.

Solomon RL, Corbit JD. An opponent-process theory of motivation: I. Temporal dynamics of affect. Psychol Rev 81: 119-45, 1974.

Steinberg H, McMillan TM. Lithium and reduced consumption of drugs of abuse. pp 61-8 in Lithium in Medical Practice, ed FN Johnson, S Johnson. MTP Press, Lancaster, 1978.

Stern P, Lekovic D, Przic R, Casparovic I. Beitrag zur Pharmakologie des Lithiums. Arch Int Pharmacodyn Ther 133: 58-65, 1961.

Taukulis HK, Revusky SH. Odor as a conditioned inhibitor: Applicability of the Rescorla-Wagner model to feeding behavior. Learn Motiv 6: 11-27, 1975.

Thomsen K, Olesen OV. Lithium-induced acute renal failure in the rat. Toxicol Appl Pharmacol 45: 155-61, 1978.

Venkatakrishna-Bhatt H, Bures J. Paradoxical sleep deprivation and LiCl poisoning: Effects on the sleep-wakefulness pattern in rats. Physiol Bohemoslav 27: 282-3, 1978a.

Venkatakrishna-Bhatt H, Bures J. Electrophysiolo-gical changes induced by paradoxical sleep deprivation and lithium chloride poisoning in rats.   Brain Res 152: 97-103, 1978b.

Venkatakrishna-Bhatt H, Bures J, Buresova O. Differential effect of paradoxical sleep deprivation on acquisition and retrieval of conditioned taste aversion in rats.   Physiol Behav 20: 101-7, 1978.

Wagner AR. Priming in STM: An information processing mechanism for self-generated or retrieval-generated depression of performance.   In Habituation: Perspectives from Child Development, Animal Behavior, and Neurophysiology, ed TJ Tighe, RN Leaton.   NJ Hillsdale, Erlbaum, 1976.

Watts ME, Mark RR. Drug inhibition of memory formation in chickens.   Short-term memory. Proc R Soc Lond (Biol) 178: 455-64, 1970.

Weeden RP. Lead enhancement of lithium-induced polydipsia.   Science 205: 725-6, 1979.

Weischer M-L. Einfluss von Lithium und Rubidium auf Neugierverhalten und lokomotorische Aktivität isoliert gehaltener männlicher Mäuse. Psychopharmacology 61: 263-6, 1979.

Westbrook RF, Hardy WT, Faulks I. The effects of lithium upon drinking in the pigeon and the rat.   Physiol Behav 23: 861-4, 1979.

White N, Sklar L, Amit Z. The reinforcing action of morphine and its paradoxical side effect. Psychopharmacology 52: 63-6, 1977.

Wielosz M. Lithium, stimulants and behavior. pp 69-82 in Origin, Prevention and Treatment of Affective Disorders, ed M Schou, E Strömgren. Academic Press, London 1979.

Wielosz M, Kleinrok Z. Lithium-induced head-twitches in rats.   J. Pharm Pharmacol 31: 410-1, 1979.

Yamada K, Furakawa T. Serotonergic function in

mouse head twitches induced by lithium and reserpine. Psychopharmacology 61: 255-60, 1979.

Zakusov VV, Lyubimov BI, Yavorskii AN, Fokin VI. Effect of lithium chloride on ethanol consumption by rats. Byull Eksp Biol Med 83: 693-6, 1977.

Zhukov VN. Effect of neurotropic drugs on sleep disturbance caused by electrical stimulation of the hypothalamus in cats. Bull Exp Biol Med 83: 667-70, 1977.

Zilberman Y, Kapitulnik J, Feuerstein G, Lichtenberg D. Effect of prolonged lithium treatment in the water consumption and lithium content of rats. Pharmacol Res Commun 11: 467-74, 1979.

AUTHOR INDEX